ROBOTIC TELESCOPES: CURRENT CAPABILITIES, PRESENT DEVELOPMENTS, AND FUTURE PROSPECTS FOR AUTOMATED ASTRONOMY

A SERIES OF BOOKS ON RECENT DEVELOPMENTS IN ASTRONOMY AND ASTROPHYSICS

**A.S.P. CONFERENCE SERIES
PUBLICATIONS COMMITTEE**

Dr. Sallie L. Baliunas, Chair
Dr. John P. Huchra
Dr. Roberta M. Humphreys
Dr. Catherine A. Pilachowski

© Copyright 1995 Astronomical Society of the Pacific
390 Ashton Avenue, San Francisco, California 94112

All rights reserved

Printed by BookCrafters, Inc.

First published 1995

Library of Congress Catalog Card Number: 95-76336
ISBN 0-937707-98-8

D. Harold McNamara, Managing Editor of Conference Series
408 ESC Brigham Young University
Provo, UT 84602
801-378-2298

A SERIES OF BOOKS ON RECENT DEVELOPMENTS IN ASTRONOMY AND ASTROPHYSICS

Vol. 1-Progress and Opportunities in Southern Hemisphere Optical Astronomy: The CTIO 25th Anniversary Symposium
ed. V. M. Blanco and M. M. Phillips   ISBN 0-937707-18-X

Vol. 2-Proceedings of a Workshop on Optical Surveys for Quasars
ed. P. S. Osmer, A. C. Porter, R. F. Green, and C. B. Foltz   ISBN 0-937707-19-8

Vol. 3-Fiber Optics in Astronomy
ed. S. C. Barden   ISBN 0-937707-20-1

Vol. 4-The Extragalactic Distance Scale: Proceedings of the ASP 100th Anniversary Symposium
ed. S. van den Bergh and C. J. Pritchet   ISBN 0-937707-21-X

Vol. 5-The Minnesota Lectures on Clusters of Galaxies and Large-Scale Structure
ed. J. M. Dickey   ISBN 0-937707-22-8

Vol. 6-Synthesis Imaging in Radio Astronomy: A Collection of Lectures from the Third NRAO Synthesis Imaging Summer School
ed. R. A. Perley, F. R. Schwab, and A. H. Bridle   ISBN 0-937707-23-6

Vol. 7-Properties of Hot Luminous Stars: Boulder-Munich Workshop
ed. C. D. Garmany   ISBN 0-937707-24-4

Vol. 8-CCDs in Astronomy
ed. G. H. Jacoby   ISBN 0-937707-25-2

Vol. 9-Cool Stars, Stellar Systems, and the Sun. Sixth Cambridge Workshop
ed. G. Wallerstein   ISBN 0-937707-27-9

Vol. 10-The Evolution of the Universe of Galaxies. The Edwin Hubble Centennial Symposium
ed. R. G. Kron   ISBN 0-937707-28-7

Vol. 11-Confrontation Between Stellar Pulsation and Evolution
ed. C. Cacciari and G. Clementini   ISBN 0-937707-30-9

Vol. 12-The Evolution of the Interstellar Medium
ed. L. Blitz   ISBN 0-937707-31-7

Vol. 13-The Formation and Evolution of Star Clusters
ed. K. Janes   ISBN 0-937707-32-5

Vol. 14-Astrophysics with Infrared Arrays
ed. R. Elston   ISBN 0-937707-33-3

Vol. 15-Large-Scale Structures and Peculiar Motions in the Universe
ed. D. W. Latham and L. A. N. da Costa   ISBN 0-937707-34-1

Vol. 16-Atoms, Ions and Molecules: New Results in Spectral Line Astrophysics
ed. A. D. Haschick and P. T. P. Ho   ISBN 0-937707-35-X

Vol. 17-Light Pollution, Radio Interference, and Space Debris
ed. D. L. Crawford   ISBN 0-937707-36-8

Vol. 18-The Interpretation of Modern Synthesis Observations of Spiral Galaxies
ed. N. Duric and P. C. Crane   ISBN 0-937707-37-6

Vol. 19-Radio Interferometry: Theory, Techniques, and Application, IAU Colloquium 131
ed. T. J. Cornwell and R. A. Perley   ISBN 0-937707-38-4

Vol. 20-Frontiers of Stellar Evolution, celebrating the 50th Anniversary of McDonald Observatory
ed. D. L. Lambert   ISBN 0-937707-39-2

Vol. 21-The Space Distribution of Quasars
ed. D. Crampton   ISBN 0-937707-40-6

Vol. 22-Nonisotropic and Variable Outflows from Stars
ed. L. Drissen, C. Leitherer, and A. Nota   ISBN 0-937707-41-4

Vol. 23-Astronomical CCD Observing and Reduction Techniques
ed. S. B. Howell   ISBN 0-937707-42-4

Vol. 24-Cosmology and Large-Scale Structure in the Universe
ed. R. R. de Carvalho   ISBN 0-937707-43-0

Vol. 25-Astronomical Data Analysis Software and Systems I
ed. D. M. Worrall, C. Biemesderfer, and J. Barnes   ISBN 0-937707-44-9

Vol. 26-Cool Stars, Stellar Systems, and the Sun, Seventh Cambridge Workshop
ed. M. S. Giampapa and J. A. Bookbinder   ISBN 0-937707-45-7

Vol. 27-The Solar Cycle
ed. K. L. Harvey   ISBN 0-937707-46-5

Vol. 28-Automated Telescopes for Photometry and Imaging
ed. S. J. Adelman, R. J. Dukes, Jr., and C. J. Adelman ISBN 0-937707-47-3

Vol. 29-Workshop on Cataclysmic Variable Stars
ed. N. Vogt   ISBN 0-937707-48-1

Vol. 30-Variable Stars and Galaxies, in honor of M. S. Feast on his retirement
ed. B. Warner   ISBN 0-937707-49-X

Vol. 31-Relationships Between Active Galactic Nuclei and Starburst Galaxies
ed. A. V. Filippenko   ISBN 0-937707-50-3

Vol. 32-Complementary Approaches to Double and Multiple Star Research, IAU Collouquium 135
ed. H. A. McAlister and W. I. Hartkopf   ISBN 0-937707-51-1

Vol. 33-Research Amateur Astronomy
ed. S. J. Edberg   ISBN 0-937707-52-X

Vol. 34-Robotic Telescopes in the 1990s
ed. A. V. Filippenko   ISBN 0-937707-53-8

Vol. 35-Massive Stars: Their Lives in the Interstellar Medium
ed. J. P. Cassinelli and E. B. Churchwell   ISBN 0-937707-54-6

Vol. 36-Planets and Pulsars
ed. J. A. Phillips, S. E. Thorsett, and S. R. Kulkarni ISBN 0-937707-55-4

Vol. 37-Fiber Optics in Astronomy II
ed. P. M. Gray   ISBN 0-937707-56-2

Vol. 38-New Frontiers in Binary Star Research
ed. K. C. Leung and I. S. Nha   ISBN 0-937707-57-0

Vol. 39-The Minnesota Lectures on the Structure and Dynamics of the Milky Way
ed. Roberta M. Humphreys   ISBN 0-937707-58-9

Vol. 40-Inside the Stars, IAU Colloquium 137
ed. Werner W. Weiss and Annie Baglin   ISBN 0-937707-59-7

Vol. 41-Astronomical Infrared Spectroscopy: Future Observational Directions
ed. Sun Kwok   ISBN 0-937707-60-0

Vol. 42-GONG 1992: Seismic Investigation of the Sun and Stars
ed. Timothy M. Brown   ISBN 0-937707-61-9

Vol. 43-Sky Surveys: Protostars to Protogalaxies
ed. B. T. Soifer   ISBN 0-937707-62-7

Vol. 44-Peculiar Versus Normal Phenomena in A-Type and Related Stars
ed. M. M. Dworetsky, F. Castelli, and R. Faraggiana   ISBN 0-937707-63-5

Vol. 45-Luminous High-Latitude Stars
ed. D. D. Sasselov   ISBN 0-937707-64-3

Vol. 46-The Magnetic and Velocity Fields of Solar Active Regions, IAU Colloquium 141
ed. H. Zirin, G. Ai, and H. Wang   ISBN 0-937707-65-1

Vol. 47-Third Decinnial US-USSR Conference on SETI
ed. G. Seth Shostak   ISBN 0-937707-66-X

Vol. 48-The Globular Cluster-Galaxy Connection
ed. Graeme H. Smith and Jean P. Brodie   ISBN 0-937707-67-8

Vol. 49-Galaxy Evolution: The Milky Way Perspective
ed. Steven R. Majewski   ISBN 0-937707-68-6

Vol. 50-Structure and Dynamics of Globular Clusters
ed. S. G. Djorgovski and G. Meylan   ISBN 0-937707-69-4

Vol. 51-Observational Cosmology
ed. G. Chincarini, A. Iovino, T. Maccacaro, and D. Maccagni   ISBN 0-937707-70-8

Vol. 52-Astronomical Data Analysis Software and Systems II
ed. R. J. Hanisch, J. V. Brissenden, and Jeannette Barnes   ISBN 0-937707-71-6

Vol. 53-Blue Stragglers
ed. Rex A. Saffer   ISBN 0-937707-72-4

Vol. 54-The First Stromlo Symposium: The Physics of Active Galaxies
ed. Geoffrey V. Bicknell, Michael A. Dopita, and Peter J. Quinn   ISBN 0-937707-73-2

Vol. 55-Optical Astronomy from the Earth and Moon
ed. Diane M. Pyper and Ronald J. Angione   ISBN 0-937707-74-0

Vol. 56-Interacting Binary Stars
ed. Allen W. Shafter   ISBN 0-937707-75-9

Vol. 57-Stellar and Circumstellar Astrophysics
ed. George Wallerstein and Alberto Noriega-Crespo   ISBN 0-937707-76-7

Vol. 58-The First Symposium on the Infrared Cirrus and Diffuse Interstellar Clouds
ed. Roc M. Cutri and William B. Latter   ISBN 0-937707-77-5

Vol. 59-Astronomy with Millimeter and Submillimeter Wave Interferometry
ed. M. Ishiguro and Wm. J. Welch   ISBN 0-937707-78-3

Vol. 60-The MK Process at 50 Years: A Powerful Tool for Astrophysical Insight
ed. C. J. Corbally, R. O. Gray, and R. F. Garrison   ISBN 0-937707-79-1

Vol. 61-Astronomical Data Analysis Software and Systems III
ed. Dennis R. Crabtree, R. J. Hanisch, and Jeannette Barnes   ISBN 0-937707-80-5

Vol. 62-The Nature and Evolutionary Status of Herbig Ae / Be Stars
ed. P. S. Thé, M. R. Pérez, and E. P. J. van den Heuvel   ISBN 0-937707-81-3

Vol. 63-Seventy-Five Years of Hirayama Asteroid Families: The role of Collisions in the Solar System History
ed. R. Binzel, Y. Kozai, and T. Hirayama   ISBN 0-937707-82-1

Vol. 64-Cool Stars, Stellar Systems, and the Sun, Eighth Cambridge Workshop
ed. Jean-Pierre Caillault   ISBN 0-937707-83-X

Vol. 65-Clouds, Cores, and Low Mass Stars
ed. Dan P. Clemens and Richard Barvainis   ISBN 0-937707-84-8

Vol. 66- Physics of the Gaseous and Stellar Disks of the Galaxy
ed. Ivan R. King   ISBN 0-937707-85-6

Vol. 67-Unveiling Large-Scale Structures Behind the Milky Way
ed. C. Balkowski and R. C. Kraan-Korteweg   ISBN 0-937707-86-4

Vol. 68-Solar Active Region Evolution: Comparing Models with Observations
ed. K. S. Balasubramaniam and George W. Simon   ISBN 0-937707-87-2

Vol. 69-Reverberation Mapping of the Broad-Line Region in Active Galactic Nuclei
ed. P. M. Gondhalekar, K. Horne, and B. M. Peterson   ISBN 0-937707-88-0

Vol. 70-Groups of Galaxies
ed. Otto G. Richter and Kirk Borne   ISBN 0-937707-89-9

Vol. 71-Tridimensional Optical Spectroscopic Methods in Astrophysics
ed. G. Comte and M. Marcelin   ISBN 0-937707-90-2

Vol. 72-Millisecond Pulsars—A Decade of Surprise, ed. A. A. Fruchter
M. Tavani, and D. C. Backer   ISBN 0-937707-91-0

Vol. 73-Airborne Astronomy Symposium on the Galactic Ecosystem: From Gas to Stars to Dust
ed. M. R. Haas, J. A. Davidson, and E. F. Erickson   ISBN 0-937707-92-9

Vol. 74-Progress in the Search for Extraterrestrial Life,
ed. G. Seth Shostak   ISBN 0-937707-93-7

Vol. 75-Multi-Feed Systems for Radio Telescopes
ed. D. T. Emerson and J. M. Payne   ISBN 0-937707-94-5

Vol. 76-GONG '94: Helio- and Astero-Seismology from the Earth and Space
ed. Roger K. Ulrich, Edward J. Rhodes, Jr., and Werner Däppen   ISBN 0-937707-95-3

Vol. 77-Astronomical Data Analysis Software and Systems IV
ed. R. A. Shaw, H. E. Payne, and J. J. E. Hayes   ISBN 0-937707-96-1

Vol. 78-Astrophysical Applications of Powerful New Databases
ed. S. J. Adelman and W. L. Wiese   ISBN 0-937707-97-X

Vol. 79-Robotic Telescopes: Current Capabilities, Present Developments, and Future Prospects for Automated Astronomy
ed. Gregory W. Henry and Joel A. Eaton   ISBN 0-937707-98-8

Inquiries concerning these volumes should be directed to the:
Astronomical Society of the Pacific
CONFERENCE SERIES
390 Ashton Avenue
San Francisco, CA 94112-1722
415-337-1100
e-mail asp @ stars.sfsu.edu

ASTRONOMICAL SOCIETY OF THE PACIFIC
CONFERENCE SERIES

Volume 79

ROBOTIC TELESCOPES: CURRENT CAPABILITIES,
PRESENT DEVELOPMENTS, AND FUTURE
PROSPECTS FOR AUTOMATED ASTRONOMY

Proceedings of a Symposium held as part of the 106th
Annual Meeting of the Astronomical Society of the Pacific
Flagstaff, Arizona
28-30 June 1994

Edited by
Gregory W. Henry and Joel A. Eaton
Tennessee State University
Nashville, Tennessee 37203 USA

TABLE OF CONTENTS

Preface xii

Registered Participants xiv

Conference Summary: Looking Inward
   Virginia Trimble 1

PART I - CURRENT CAPABILITIES

The Phoenix 10 Automatic Photometric Telescope: An Update
   Michael A. Seeds 11

APT Observations of Small-Amplitude Red Variables: Science and Psychology
   John R. Percy, Winnie Au 17

The Four College Consortium APT: The First Four Years
   Robert J. Dukes, Jr., William R. Kubinec, Harold L. Nations, Saul J. Adelman,
   Diane P. Smith, Edward F. Guinan, George P. McCook 20

The Fairborn/TSU Robotic Telescope Operations Model
   Gregory W. Henry 37

The Development of Precision Robotic Photometry
   Gregory W. Henry 44

Variable Stars in the Hertzprung-Russell Diagram
   Douglas S. Hall 65

Integrated Telescope and CCD Control on a PC
   Norman L. Markworth 81

The Berkeley Automatic Imaging Telescope: An Update
   Richard R. Treffers, Alexei V. Filippenko, Schuyler D. Van Dyk,
   Young Paik, Michael W. Richmond 86

The Micro-Observatory: An Automated Telescope for Education
   P. Steven Leiker, Philip M. Sadler, Kenneth Brecher 93

PART II - PRESENT DEVELOPMENTS

Flexible Scheduling of Automatic Telescopes over the Internet
Mark Drummond, John Bresina, Will Edgington, Keith Swanson
Greg Henry, Ellen Drascher 101

Automated Telescope Monitoring and Diagnosis
Christine M. Monahan, F. A. Patterson-Hine, David L. Iverson 120

Diagnostic Algorithms and Health Monitoring of Automatic Telescopes
Kenneth M. Valentine, Michael A. Donahue, Robert W. Valentine 129

Operations Issues for the Hobby-Eberly Telescope
Phillip W. Kelton, Mark E. Cornell 136

Automating Mission Scheduling for Space-Based Observatories
Nicola Muscettola, Barney Pell, Othar Hansson, Sunil Mohan 148

Scheduling the IUE Satellite with Constraint Logic
Bruce McCollum, Mark Graves 167

Planning and Scheduling for the Hubble Space Telescope
Glenn E. Miller 173

Advances in Autonomous Operations for the EUVE Science Payload and Spacecraft
T. Morgan, R. F. Malina 184

The South Pole Infrared Explorer (SPIREX): Near Infrared Astronomy
at the South Pole
Bernard J. Rauscher, Mark Hereld, Hien Nguyen, Scott Severson 195

Antarctic Muon and Neutrino Detector AMANDA: First Data and Outlook
John Lynch 205

PART III - FUTURE PROSPECTS

Small Telescopes and NOAO's New Venture
David L. Crawford 223

The Rationale for an Automatic Spectroscopic Telescope
Joel A. Eaton 226

A Proposal for a Fiber-Fed Échelle Spectrograph for a Southern-Hemisphere
Robotic Telescope
J. B. Hearnshaw 233

Network of Oriental Robotic Telescopes
  Francois Rene Querci, Monique Querci, Samir Kadiri, Zouhair Benkhaldoun . . . 239

Astronomy from the Moon
  Jack O. Burns 242

A Practical and Affordable Telescope for the Moon
  Peter C. Chen, R. J. Oliversen, H. Hojaji, K. B. Ma, M. Lamb, W. K. Chu 252

Author Index 269

PREFACE

A symposium on Robotic Telescopes and Automated Astronomy was held in conjunction with the 106th Annual Meeting of the Astronomical Society of the Pacific during June 28-30, 1994 in Flagstaff, Arizona. Approximately 60 people attended the symposium and 39 presented talks. This volume holds the 26 papers submitted for publication in the proceedings of the meeting.

Our own interest in automated astronomy was piqued early on by long photometric runs at Kitt Peak to observe variable stars. One of us in particular recalls how he could barely raise his arm past shoulder height after three nights of writing data on a strip chart. We both remember the mind-numbing repetition of this work and the inevitable fantasy that the National Observatory would automate at least some of the steps involved. Things have changed considerably since those days, mostly for the better. The National Observatory never did automate our manual photometry but, instead, closed our telescope. Computers got a lot cheaper and more capable, and amateur astronomers stepped into the breach with robotic telescopes that could find stars and measure their brightnesses without needing a human observer. These machines have grown in size and sophistication to become fully professional instruments better than anything Kitt Peak could have offered us. This symposium is one of a long series of meetings chronicling and advancing that transformation.

One of the highlights of the meeting was the Distinguished Service Award to Lou Boyd who built the first truly successful automatic photometric telescope. Lou's Fairborn Observatory now has eight of these telescopes operating atop Mt. Hopkins in southern Arizona for various groups of astronomers. Other such APT's are operating in America, Italy, and New Zealand. Further, the trend toward automation has infected the very heart of observational astronomy--CCD imaging and spectroscopy, and we have papers on both subjects in this Proceedings.

Part I assesses the current capabilities of existing robotic telescopes and the kinds of science observations that are being made with them. Besides photometric observations, automated imaging is now becoming routine. Part II outlines some of the developments currently taking place to advance automated astronomy. The majority of papers in this section discuss automated scheduling of ground-based and space-based telescopes. The advantages of conquering the South Pole for automated observations are also discussed in this section. Part III looks forward to the development of automated spectroscopy, networks of robotic telescopes, and automated observations from the moon.

This proceedings contains most of the new results presented at the symposium. However, several excellent talks have not made it into print. These include Lou Boyd (Ten Years of Automated Photometry), Kent Honeycutt (Automated CCD Photometry of Cataclysmic Variables - see eg., ApJ, 425, 835, 1994), Wes Lockwood (Comparison Between Human and Robotic Photometry), Charles Townes (Berkeley Interferometer on Mt. Wilson), Theo Brummelaar (CHARA Interferometer Array), Russ Genet (Multiple Telescope Robotic Observatories), Larry Ramsey (Hobby-Eberly Telescope), John Lynch (Polar Aeronomy and Astrophysics), Al Harper (Infrared and Submillimeter Work at South Pole), Tom Duvall (Helioseismology of the Sun from South Pole), John Bally (South Pole Advanced Telescope Project), and Mark Colavita (Interferometry from the Moon).

No conference is a success without the tireless work of its organizers. This one benefitted especially from the efforts of Robert Havlen, executive director of the Astronomical Society of the Pacific. We also thank Russ Genet (ASP President) and Mark Drummond (Recom/NASA Ames) for their help in organizing the scientific sessions. We are especially grateful to Tennessee State University and the College of Charleston for help in defraying some of the costs associated with the symposium.

Gregory W. Henry
Joel A. Eaton
Tennessee State University
Nashville, Tennessee

REGISTERED PARTICIPANTS

Ronald Angione	San Diego State University
John Bally	University of Colorado
Morrie Barembaum	Whittier, CA
Robert Bell	Lake Arrowhead, CA
Jean Pierre Berger	Observatoire de St. Mihil
William Blankley	California Lutheran University
Louis Boyd	Fairborn Observatory
John Bresina	NASA Ames/Recom Technologies
Theo Brummelaar	Georgia State University
Jack Burns	New Mexico State University
Alan Chen	Columbia, MD
Peter Chen	NASA Goddard Space Flight Center
Lester Clendenning	Humbolt State University
Mark Colavita	Jet Propulsion Laboratory
David Crawford	NOAO-Kitt Peak
Ron Dantowitz	Hayden Planetarium
James DeVeny	NOAO-Kitt Peak
Gary Dowdle	San Antonio, TX
Mark Drummond	NASA Ames/Recom Technologies
Robert Dukes, Jr.	The College of Charleston
Thomas Duvall	NASA Goddard Space Flight Center
Joel Eaton	Tennessee State University
William Edgington	NASA Ames/Recom Technologies
Donald Epand	Fairborn Observatory
Henry Funston	Hereford, AZ
R. H. Garstang	University of Colorado
Russell Genet	Mesa, AZ
Douglas Hall	Vanderbilt University
Jeffrey Hall	Lowell Observatory
Al Harper	Yerkes Observatory
John Hearnshaw	University of Canterbury
Arne Henden	US Naval Observatory
Gregory Henry	Tennessee State University
Kent Honeycutt	Indiana University
Matt Hungerford	McKinelyville, CA
W. H. Jamison	Rocky Mountain College
Stuart Jefferies	National Solar Observatory
Phillip Kelton	University of Texas McDonald Observatory
Steven Leiker	Harvard-Smithsonian Center for Astrophysics
G. W. Lockwood	Lowell Observatory
John Lynch	National Science Foundation
Norman Markworth	Stephen F. Austin State University
Bruce McCollum	NASA Goddard Space Flight Center
Glenn Miller	Space Telescope Science Institute
H. R. Miller	Georgia State University
Christine Monahan	NASA Ames/Recom Technologies

Thomas Morgan	University of California, Berkeley
Nicola Muscettola	NASA Ames Research Center
Burt Nelson	San Diego State University
Elizabeth Orosz	Columbia, SC
John Percy	University of Toronto
Sigurd Peterson III	Aloha, OR
Francois Querci	Observatoire Midi-Pyrénnées
Lawrence Ramsey	Pennsylvania State University
Bernard Rauscher	The University of Chicago
Jeffrey Rosendhal	NASA Headquarters
Wallace Sargent	Palomar Observatory
Tom Sebring	University of Texas McDonald Observatory
Michael Seeds	Franklin and Marshall College
James Seevers	Adler Planetarium
Joseph Segovia	Fullerton, CA
Jerry Sherlin	National Weather Service
Joe Shields	University of Arizona Steward Observatory
Jeff Stoner	Boulder, CO
Charles Townes	University of California, Berkeley
Richard Treffers	University of California, Berkeley
Virginia Trimble	University of California, Irvine
James Wray	Sci Tech Astronomical Research

CONFERENCE SUMMARY: LOOKING INWARD

Virginia Trimble

Physics Department, University of California, Irvine, CA 92717, and Astronomy Department, University of Maryland, College Park, MD 20742

Abstract. This final presentation at the symposium attempted to provide some perspective on where we are going in the areas of hardware, software, scientific results, and interactions with the rest of the astronomical community for automated, robotic, and remotely operated telescopes on traditional and non-traditional sites.

1. Introduction

Remote, automated, and robotic observing, broadly defined, have a considerable paleolithic history, prior to the commissioning of the Phoenix 10" telescope (described by L. Boyd). The very early UV and X-ray rocket experiments did not much more than open and close the "dome" on their own, parachuting the packages down for investigators to retrieve and examine. By 1960, Sputnik I, Explorer, and Vanguard had collected and returned data automatically, while the Wisconsin X-15 rocket plane UV observations had both remote operation (from the pilot's cabin!) and automatic tracking and observing. Their Pine Bluff Observatory could operate for several days without human intervention, albeit primarily as an extinction monitor. OAO-2 collected real data (though with frequent hints from the ground on what to do next) for the four years 1968-72, and the Coralitos supernova search had automated aspects at about the same time. Prior to the HST repair mission and the SMM "fix-it" mission mentioned at the symposium, ISEE-3 (International Sun-Earth Explorer) had been reconfigured, or at least re-aimed, as ICE (International Comet Explorer) remotely in anticipation of Halley's return (which occurred on schedule, though neither NASA nor ESA can take much of the credit). And the all-time success story remains IUE, which was launched when many of the present participants had not yet received their first degrees.

Increasing our temporal magnification by about an order of magnitude permits comparison of current conditions with the state of the subject in June 1991, when ASP hosted "Robotic Telescopes in the 1990s" (ASP Conf. Vol. 34, ed. A.V. Filippenko) at Laramie, Wyoming, an even higher-elevation site than Flagstaff.

2. Hardware

The shear numbers of partly- and fully-automated telescopes are increasing rapidly, from about five in 1989 to about 30 now (L. Boyd, R. Genet, G. Henry). Photometric precision has also unambiguously improved over the period from night-to-night variations of 0.01 mag down to 0.001 mag for those installations that monitor extinction frequently (G. Henry). Long term stability is always more difficult to achieve. G. W. Lockwood's presentation indicated that real stellar drifts as small as 0.001 mag/yr could be identified by traditional observer-operated systems. It is not clear that the APTs can yet match this level.

Sizes of operating automated telescopes are also increasing, in the sense that most now being commissioned are larger than the pioneering Phoenix 10" and Fairborn 10". (The Pine Bluff was an 8".) The MACHO-search 50" (formerly the Great Melbourne Telescope!) seems to be the largest with largely automated data collection at the present time, while HST is the largest fully-remote mirror in operation. L. Ramsey and P. W. Kelton pointed out that the Hobby-Eberly (formerly spectroscopic survey) Telescope will be a 9 to 11 meter (depending on where you put your meter stick), but is intended for only partially remote or automated operation.

Guiding has become more common, permitting both longer exposure photometry and imaging, as in the Berkeley Automated Imaging Telescope (R. Treffers). There was little discussion of this or other aspects of long-term stability. The rapid sort of guiding normally called active and adaptive optics is of necessity an automated procedure, also not discussed except in the infrared context of the Berkeley Infrared Spatial Interferometer (C. H. Townes).

Interferometry is clearly the most extreme version of coordination of two or more mirrors, and the provision of delay lines indicates the intention to make operation "hands-off". The Berkeley ISI currently operates in an intermediate mode, but increasing automation is intended. The 36 Keck mirrors were designed to maintain their collective figure of revolution continuously, and generally do so. Keck I and II and the four mirrors of the ESO Very Large Array are intended for interferometric operation with a minimum of human intervention, though not in the near future. The first fully automated optical array could well be that of CHARA (center for High Angular Resolution Astronomy, T. ten Brummelaar), which, if funding were guaranteed tomorrow, could be operational by 2001 or thereabouts.

Automated coordination (though not coherence!) is intended for both GNAT (Global Network of Automated Telescopes, D. Crawford) and ORT (Oriental Robotic Telescopes, F. Querci). The closest to an operational system of this sort is the Whole Earth Telescope, which is already producing data on variable stars, but was not discussed at the symposium.

At a still looser level of linkage, R. Genet discussed multi-telescope and multi-purpose sites. Many such operations are under way (most notably Mauna Kea, already shared by at least 5 countries - US, UK, Canada, France, and Japan), partly as an economy measure and partly because excellent sites are rare. The south pole and the moon are both unique and cases where even small reductions in numbers of people make a big difference in dollars.

On a less cheerful note, automated spectroscopy is still not a reality. My 1991 summary described it as "components exist; some assembly required."

Judging from the presentations of J. Eaton and J. Hernshaw, this is still the case, though AutoScope is working on 1.5 – 2.5m class telescopes particularly suited for the purpose. The prime mover for automated radial velocity spectroscopy, seemingly an especially appropriate mating, has apparently given up on the project. A bright spot is the Sloan Digital Sky Survey, for which both hardware and software are being formulated for computerized data acquisition, processing, and storage (M. Richmond, not a participant).

In recent years, professional optical astronomers have typically not been closely involved in designing and building their own telescopes (in contrast to the radio, X-ray, Gamma-ray, and, of course, amateur optical communities). It has been suggested that Alvan G. Clark was the last astronomer to make a fundamental discovery with a telescope he had himself built (it was the companion of Sirius, the first white dwarf), and that some of modern astronomical woes may derive from this decoupling. The involvement of NOAO in industrial plans for 0.5 – 2.5m telescopes (D. Crawford) is, therefore, an encouraging development.

3. Software

Real progress has occurred in the software for scheduling remotely and automatically operated telescopes. A few years ago, standard APTs spent a very large fraction close either to the western horizon (chasing objects about to set) or to the eastern horizon (chasing ones just coming into view). The graphs shown by L. Boyd, G. Henry, and R. Dukes made clear that this is no longer typically the case. Comparable improvement has occurred for other kinds of telescopes. At the time HST was first scheduled for launch, it took 30 hours of computing time to prepare for 24 hours of observing (ominous in the extreme). The same schedule can now be assembled in 30 minutes, and redone a couple of times per day, if necessary, with an appropriate mix of Spike and SPSS for long- and short-term management (G. Miller). IUE "observing runs" are arranged with comparable efficiency, additional flexibility coming from the assignment of 8-hour blocks to particular projects (B. McCollum). Thus a bright supernova can be targeted during a particular shift previously assigned to K giant winds and the wind project given a later shift without having to displace anything else. And the 7 second scheduling time per night of the Berkeley Automatic Imaging Telescope (R. Treffers) allows its plans to be "rethought" after every observation, even though scientific priorities, as well as logistical criteria, are considered by the program.

Another piece of good news is that, although grabbing back manual control in real time is not generally available (M. Drummund) post-observation human intervention can effectively identify unexpected, as well as expected, data patterns. The OGLE lensing search, for instance, will (though not discussed at the symposium) yield a complete catalog of all variable point sources observed.

On the "better luck next time" side N. Markworth and N. Muscettola warned of built in glitches, like a control system that turns off the guider as soon as a target is acquired, or two subsystems that need to talk to each other to make sure the operation of locking the pointing device on to a target is coordinated with the target being visible!

A large fraction of the symposium was devoted to ongoing work on scheduling and other kinds of software (G. Henry, L. Boyd, J. L. Bresina, C. Monahan – the other female speaker, and T. H. Morgan), including the importance of standardization (D. Crawford). It is perhaps best not to mention which of these various talks made me wonder about the correct name for a software system that is the analog of hardware designed by Rube Goldberg, but there is perhaps some tendency to continue to use and adapt inappropriate packages that one is comfortable with. The issues are in any case important. While most installations do not have quite the $20/second cost of HST "open-slit" time, actually using 20% more of the time available to a given telescope is the equivalent of building 20% more of them.

4. Using your APT (etc.)

4.1. Kinds of Uses

Monitoring of weather conditions and atmospheric transparency have been part of most APT operations since the times of Pine Bluff and the Phoenix 10". It is not necessary to close down, short of actual rain, just because things get hazy. The TSU telescope keeps going as long as it can, interspersing quality control observations. A human observer looks at these later and decides which data points to keep. And, of course, a robotic telescope is less likely than a night assistant to object to re-opening at 4 AM if conditions improve. A system need not be very complicated to be useful. B.A.I.T. simply "assumes" that if it fails to acquire 6 targets in a row, clouds are in the way. It waits for an hour and then tries again.

Following the behavior of known interesting objects is the commonest current APT task, like the Be stars and multi-mode Cepheids studied by R. Dukes and R. K. Honeycutt's cataclysmic variables. The latter noted that the process (especially if all your program objects are the same type) is likely to yield nothing informative for ages and then a sudden deluge of product – a lot like growing zucchini.

Searches for known classes of objects have been successfully automated, including supernovae and gravitational lenses. APT data bases have not yet been much used for searching for new classes of objects, in the way that the IRAS and ROSAT catalogs are used, and this would seem to require human intervention at some stage.

Finally, many installations are intended at least partially for instructional purposes, including the Harvard Micro Observatory (S. Leiker) and the AutoScope imaging telescope recently installed at one of my two institutions (UCI). Four or five students are currently involved in debugging the hard- and software (we bought a fixer-upper at cut rate price), and this may well be the most educational part of the experience.

4.2. Scientific Results

Many of the projects under discussion three years ago have begun to produce interesting astronomy, including the Whole Earth Telescope, OGLE, MACHO, and EROS. Others are operating, including the Explosive Transient Camera and Rapidly Moving Telescope (intended to catch gamma ray bursts with their photons down). None of these was heard from at the present symposium.

Among the projects reporting, Townes and the Berkeley ISI have now measured enough infrared angular diameters at enough times to conclude that some evolved stars produce dust continuously, other episodically, and that the latter tend to have longer pulsation periods (normally meaning lower surface temperature and more evolved state).

Extended, reliable light curves, are of course the raw materials for classifying variable stars in terms of the underlying physical mechanism (as presented by D. S. Hall), which then leads with luck, to classification by evolutionary phase and initial star or system properties. A particularly interesting case is Honeycutt's subset of CVs that seem to cycle between a bright, relatively steady, nova-like state, and a fainter, oscillatory dwarf-nova-like state. He suggests that the former, longer-term change arises in changes in the amount of material being contributed by the donor star (an instability first suggested by Bohdan Paczyski and strongly advocated by Geoff Bath) and the latter, rapid changes from an instability in viscosity (and so accretion rate) within the disk itself (first suggested by Osaki and advocated by Paczyski).

Even the sun is a variable star if you look hard enough (at its rotation period, owing to spots, and through its 11 year cycle, for more complicated reasons.) G. W. Lockwood has been monitoring solar analog stars for many years and has concluded that, while we are fairly normal in terms of cycle length and chromospheric emission for our rotation speed and age, the solar level of brightness fluctuation is anomalously low, by a factor of about three, when normalized to H & K emission, age, etc. The reason for this is not known, but it must have implications for long term climate change on earth.

5. Political issues

5.1. The External Image of the Field

As R. Dukes noted, APT users began publishing their more important results in standard journals (especially PASP) about the time I recommended this in 1991. Post hoc non ergo propter hoc! This seems a step in the right direction. And, when the first events from EROS, MACHO, and OGLE preprinted across the skies, they were doubted, but for reasons quite disjoint from the (varying) degrees of automation of the searches. This represents a significant advance, along the same lines as being able to say that you have found a cure for a particular kind of cancer when most of the people who have it survive long enough to die of something else.

5.2. Archiving

Archiving was given generic high priority in the Bahcall report as well as at previous APT symposia and at an IAU Symposium last summer on wide field imaging. So far, only NASA and ESA seem to have made much progress. Their experience has been that it costs about 10% as much as you spent acquiring the data to store it in a form useful to most astronomers (the catch is that storage is cheap, retrieval expensive, and documentation very expensive). A participant mentioned that preliminary indications from the Sloan Digital Sky Survey (for which both cameras and programs are now being assembled) is that 10% remains a good number when your telescope sits still.

5.3. Cooperation and the User Pool

Multi-user automated telescopes are quite common. We heard specifically from the Four College (Dukes), BAIT (Treffers), and Rent-A-Star (M. Seed) programs, none of which seem to be adversely affected by the horses pulling the sled in different directions. Dukes noted that ATIS scheduling for multiple users was already almost too efficient – they all get back more data than they have time to deal with!

People from within and without the APT community continue to worry about a generation of new astronomers coming along who are unable either to make best use of their data or to understand its limitations because their experience of the telescopes, detectors, and programs used to collect it is so limited. I share this concern, but suspect that it may all come out all right as long as all systems break down thoroughly once per generation (5 years?) and have to be taken apart and fixed. It is vital that the occurrence of the breakdown be unambiguously clear, and hardware tends to be more user friendly in this respect than software. Notoriously, we all keep twiddling the dials on our program until we like the answer, and the hardware equivalent is much less common – though folklore preserves a story of Alvan Clark trying valiantly to polish away the companion of Sirius, having taken it for an internal reflection.

5.4. What to Automate and Why

The two obvious drivers for various forms of automation are higher quality and lower cost. J. Percy suggested that students should observe the variable stars and APTs the constant ones, so that boredom would not set in and affect reliability (or the probability of the students finishing the course!). Digital plate blinkers are not always more accurate than biomechanical ones (the "false positive" rate, particularly, is high), but they are a lot faster, and this matters when the telescope is running continuously.

Cost is much lower even for accessible sites, a factor of ten per observation according to an estimate from the TSU group. No dollar estimates were given for the South Pole, but most of the cargo flights in carry stuff for the observers, not for the observatories, including the Spirex search for a dark 2.35 window (B. Rauscher), the helioseismology program (T. Duvall), and the (probably best-known) IR and sub-mm programs AST/RO, Cobra, and so forth (D. A. Harper).

At some level of inaccessibility, automation becomes essential. Among the various polar programs discussed by J. Lynch, the dividing line occurred for multiple-week balloon flights. In the early days of scientific ballooning, scientists normally accompanied their packages (e.g. Hess's demonstration that cosmic rays come from up above, not from down below), but flights were shorter and astronomers seem to have been made of steel in the days when telescopes were made of wood.

The Apollo science experiments were deployed and sometimes operated by people. This is not going to happen on the moon again in the next decade or, perhaps, at any time in the lifetime of anyone who was at the symposium. Thus the various possible lunar telescopes and observatories discussed by J. Burns, M. Colavita, and P. Chen were all self-deploying and fully operated by remote and/or automatic methods.

Thus one can contemplate hands-off over a whole range of time scales, from between servicing missions (one per night for Spirex, to one per five years for HST), through the time line after construction, on out of the entire assembly and deployment process (Colavita).

I had a qualitative impression that the lunar programs under discussion are considerably less ambitious than those of a few years ago and that this reflects realistic evaluation of what might conceivably be done with existing dollars and launch vehicles.

5.5. Conferences as the Pulse of a Field

I have participated in, and so have proceedings volumes from three previous meetings in this series – 1989 (Tucson), 1990 (Boston), and 1991 (Laramie). The numbers of invited and contributed papers were, respectively, 40, 53, 35, and 38 here in Flagstaff. The impression, therefore, is of subdiscipline that is still not making enough contact with the rest of the astronomical community to draw many of them into participation.

As the shadows of 1939 deepened, a British general is supposed to have said that there would always be a place in battle for the well-bred horse. He seems to have been wrong (though there are more horses, nearly all recreationally employed, in the US now than there were when they worked). But I both expect and hope that there will always be a place in astronomy for the well-bred photometrist – not to mention the ill-bred summarizer.

Acknowledgments. It is the traditional privilege of the last speaker at events like these to thank those who have made the gathering possible. First, our local hosts from Lowell Observatory, Northern Arizona University, the Conconino Astronomers, and NAU Astronomy Club, chaired by Larry Wasserman (Lowell). Second, Robert Havlen, Lonny Baker, and their associates at ASP for managing the whole sweep of the Annual meeting. Third, the scientific organizers and editors of the present symposium and its proceedings, Russ Genet, Greg Henry, and Mark Drummond. Russ has sworn that this is his last APT symposium (at least as an organizer), and we wish him well in his new endeavors. Finally, I would like to say a word in praise of the usually-maligned, extraordinarily complicated network of automated scheduling systems that enabled nearly all participants to have valid airplane reservations and on-time arrivals. Some things are even harder to arrange than observing runs.

PART I

CURRENT CAPABILITIES

THE PHOENIX 10 AUTOMATIC PHOTOMETRIC TELESCOPE: AN UPDATE

MICHAEL A. SEEDS
Astronomy Program, Franklin and Marshall College, Lancaster, PA 17604-3003, M_SEEDS@ACAD.FANDM.EDU

ABSTRACT The Phoenix 10 telescope has been in nearly continuous operation for over 10 years as a Rent-A-Star telescope providing UBV differential photometry to astronomers around the world. It is being upgraded during the summer of 1994, which should improve its efficiency and accuracy. With the precision of robotic telescopes continuing to improve, an improved set of bright standard stars is needed.

INTRODUCTION

The Phoenix 10 Automatic Photometric Telescope (APT) was built by Louis Boyd in Phoenix, Arizona. It is a 10-inch Newtonian with a pulse counting photometer using a Hammamatsu 1P21 photomultiplier and Johnson UBV filters (Boyd, Genet, and Hall 1984). It made its first night of robotic observations in October 1983 from Lou Boyd's back yard, and now, over 10 years later, it continues to observe from the APT Observatory on Mt. Hopkins south of Tucson, Arizona. Operating as a Rent-A-Star telescope, it provides service observations to users under the direction of a principal astronomer.

The telescope chooses its stars from a list of about 80 variable stars. Data is reduced by the principal astronomer at the end of each quarter and distributed to roughly 20 users who pay $2.00 per observation -- thus the term Rent-A-Star. Income from the telescope goes to the Fairborn Observatory Inc., a non-profit corporation, which maintains the APT Observatory and its telescopes. The telescope is available for use by any astronomer, professional or amateur, and data from the telescope has appeared in numerous journal articles.

TELESCOPE OPERATIONS

The unit of observation on the Phoenix 10 is the "group." A group is a set of three stars, the variable, comparison and check stars, carefully selected to maximize the precision of the photometry. The telescope always observes the stars in the same order CK S C V C V C V C S CK, where CK is the check star, S is sky, C is the comparison star, and V is the variable star. Each of these observations includes integrations through the Johnson UBV filters, and thus a complete group observation includes 33 integrations plus two dark integrations.

The telescope successfully observes about 7400 groups per year and has been doing so for over 10 years. Each year it successfully finds and centers on stars over 244,000 times and makes about 260,000 integrations. In 10 years, it has observed for a total integration time of roughly 10 months.

The Phoenix 10 can observe stars between declination limits of +75 and -10. The telescope has difficulty locating and observing groups beyond these declination extremes. Stars must not be brighter than 3rd magnitude and must not be fainter than about 8.0 magnitude. The bright magnitude limit is set by the sensitivity of the photomultiplier tube, and the faint magnitude limit is set by centering. The photometer can gather enough photons for good photometry of fainter stars, but the telescope cannot locate and center on stars fainter than 8th.

Recommended integration times are 10 seconds. Tests of integration times show that 5 second integrations do not produce as high a precision as 10 second integrations, but 20 second integrations also produce lower precision. These longer integrations provide too much time for small drive errors to accumulate and degrade the centering. With 10 second integrations, a single group observation including a typical slew and search to acquire the group, takes about 9.9 minutes.

During a successful night, the telescope begins in the west and works to the east. The extremes of telescope motion have been mapped to form an observing window outside which the telescope will not go. This avoids occultation by other telescopes, parts of the building, etc. The telescope begins the evening by observing groups near the western limit of its window, and works its way eastward through the night choosing groups from its star list. Each time a group is selected from the list, the telescope uses a random number generator to decide whether to observe the group or not according to a probability set by the user. Thus not every group is observed every night. A typical 100% group will be obseved slightly more than once a night. That is, it will usually be observed once a night and occasionally twice a night. If the telescope is not over or under loaded, it will reach its eastern limit some hours before dawn and then repeat groups until new groups rise into the window.

Data is reduced as differential photometry using mean extinction and transformation constants determined during standard star nights when the telescope observes only standard stars (Seeds, 1989; Hayes, Genet, and Seeds, 1989). Typically, there will be 1 to 3 standard star nights in a quarter.

All data are reported to the user as differential magnitudes in the form variable-comparison and check-comparison. Thus small variations in extinction etc. subtract out of the differential magnitudes if the comparison star is close to the variable and similar in color. Users also receive a night report which lists parameters such as the length of the night, the number of starts, stops, etc. which allows users to edit their data according to their tastes and discard nights of questionable quality (Seeds 1994).

TELESCOPE MANAGEMENT

The Phoenix 10 is not a full ATIS telescope. That is, it does not use the Automatic Telescope Instruction Set (ATIS) (Genet and Hayes, 1989 and Boyd et al, 1993). A telescope designed for control through ATIS can be directed to observe with specific filters and diaphragms at specific times and hour angles and to include or omit specific integrations of check, comparison and variable star. The Phoenix 10 is more limited. Output to the principal astronomer is in standard ATIS output format, but the control system cannot accept ATIS instruction files. Rather the telescope always uses the same filter set and makes the same integrations on variable, comparison, and check stars. Also, it uses its own selection criteria to choose stars from its observing list. Exactly which star will be observed at any given time is up to the telescope.

Because of these limitations, the principal astronomer must look at output files to deduce how well the telescope is working. The data itself is diagnostic, and the control system keeps logs files which list telescope starts, telescope stops, group aborts, and so on. In the past, changes in these files have alerted the human astronomers to bad star coordinates, damaged filters and loose drive components.

One of the files that the principal astronomer can use is the abort list. If the telescope starts a group and as it moves among the three stars it cannot find a star, it aborts the group and goes on to the next group. Aborts can occur because of bad coordinates, passing clouds, twilight, slippage in the drive, etc. It isn't unusual for a typical group to abort a few times per quarter. In the fall of 1993, however, the abort rate more than tripled, apparently because of increasing wear in the telescope drive. Although the average number of aborts per group increased dramatically, the typical photometric accuracy did not change. The average number of observations per night decreased slightly, showing that the telescope is now wasting more time on unsuccessful groups.

THE 1994 UPGRADE

Because of the increasing abort rate, Louis Boyd is upgrading the Phoenix 10 during the summer of 1994. The upgrade is expected to include both mechanical and computer upgrades that should result in improved performance, easier maintenance and better control.

The old drives used chain and sprocket on both RA and Dec, and this resulted in a small cogging error. Integrations of 20 seconds were less accurate than integrations of 10 seconds because cogging error could move the star appreciably in the diaphragm during the time it takes to observe a group with 20 second integrations, about 15 minutes. The upgrade will include new RA and Dec drives that will eliminate the chains. The existing gear boxes are badly worn and will also be replaced. In addition, microstepping motor drivers should add to the drive precision.

A new 68000 computer will be added, and this will allow improved computer control programs. This may increase telescope efficiency, but it will certainly increase the ease of operation and maintenance.

The photometer will be upgraded with a new quartz entrance window, temperature control, and dry air flooding.

These changes should result in improved data quality, easier alignment, easier update of new groups, better quality control, simplified software support, and improved telescope efficiency.

DATA QUALITY AND STANDARD STARS

Given the proposed upgrade to the telescope and its control system, it seems appropriate to look at the data quality and standard star list to establish a baseline for comparison once the telescope is fully refurbished.

A number of users have reported that with properly selected comparison stars and with proper editing to eliminate poor nights, the photometry in V and B is good to about 0.005 magnitudes. The U band, as usual, is a bit worse. This is consistent with the principal astronomers impression of overall accuracy, but to confirm that impression, specific standard star nights were reduced with the proper mean constants, and groups that were repeated through the night were compared. This selected the best quality nights and assures the statistics is applied to stars selected for their stability. The internal standard error of a single group measurement of variable minus comparison or check minus comparison is about ±0.0047 magnitudes.

The above is an internal error, but systematic errors could be larger. This was studied by comparing observations with catalog magnitudes. Catalog magnitudes of the standard stars were used to compute O-Cs in the sense observed differential magnitudes minus catalog differential magnitudes. The average of the O-Cs on a typical standard star night was -0.0007 magnitudes, but the standard deviation of a single O-C was 0.024 magnitudes. This large scatter in the O-Cs of standard stars is shown in Figure 1.

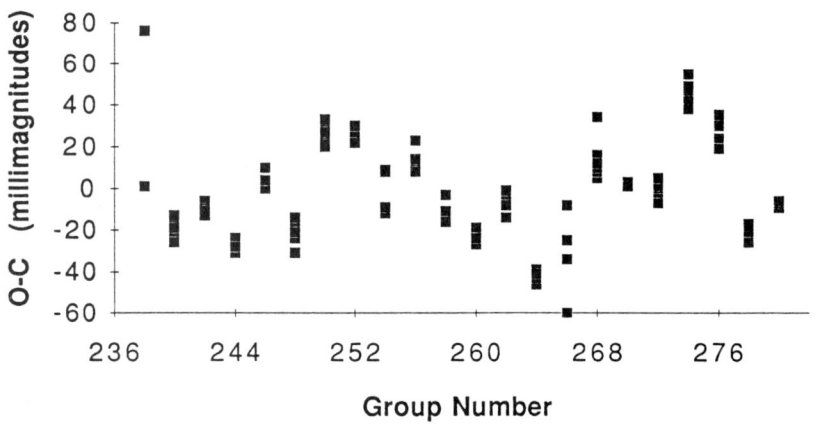

Fig. 1. Observed minus catalog differential magnitudes (in millimagnitudes) for standard star groups reveal a scatter of about ± 0.024 magnitudes. Other standard star nights show the same pattern among the groups suggesting small errors in the catalog values. Many of these standard star groups contain stars cataloged as variable.

These large errors in the standard star O-Cs seem independent of either air mass or differential air mass, and so reflect characteristics of the stars themselves. Because the O-Cs are independent of either color or separation, it seems likely that the actual standard star magnitudes being used for standard star groups on the Phoenix 10 and on some other robotic telescopes may contain errors on the order of a few hundredths of a magnitude.

Another possibility is that some of the standard stars are variable. A comparison of the standard stars with the SIMBAD data base reveals that 25 out of 48 standard stars are listed as "Variable" with 22 having NSV numbers. Two are named variable stars. The amplitudes of these stars are, when listed, on the order of a few hundredths of a magnitude, which could account for some of the scatter among the standard star O-Cs.

Standard star groups are made up like normal groups with three stars, variable, comparison, and check stars. In many standard star groups, however, the variable and check star are the same standard star, and thus the variable minus comparison and check minus comparison magnitude differences are the same. An error or a change in one star in such a group would change the two differential magnitudes by the same amount and would thus account for the systematic differences in Figure 1.

These standard magnitudes were taken from the UBVRI Standard Stars in *The Astronomical Almanac* in about 1983 and have been in use since then. While it may be disturbing to have low amplitude variables among the standard stars and to have possible catalog errors in the magnitudes, it is not surprising given the requirement that the standards be widely spread in RA and Dec and be reasonably bright. Further, these small errors in standard star magnitudes will still produce extinction and transformation coefficients of reasonable accuracy. When variable and comparison are close to each other in the sky and of similar color, even very large errors in the constants have almost no effect on the reduced differential magnitudes. Some effort should be made to select a more consistent set of bright standard stars for use by robotic telescopes, but the situation is not critical. Past data need not be rereduced, and future operations can continue with the current standards. Nevertheless, as telescopes become more precise, accuracy may be limited in part by the standard stars.

CONCLUSION

The Phoenix 10 is now entering its second decade of operation, and its upgrade during the summer of 1994 promises to improve its efficiency, accuracy, and operation. Continuing studies of the accuracy and performance of the telescope will allow an evaluation of the gains made from the upgrade. Future efforts should be made to find a more consistent set of standard stars for use by small robotic telescopes.

REFERENCES

Boyd, L. J., Epand, D., Bresina, J., Drummond, M, Swanson, K. Crawford, D. L., Genet, D. R., Genet, R. M.,Henry, G. W., McCook, G. P.,

Neely, W. Schmidtke, P., Smith, D. P., and Trueblood, M. 1993, I.A.P.P.P. Communiction No. 52, 23.
Boyd, L. J., Genet, R. M., and Hall, D. S. 1984, I.A.P.P.P. Communiction No. 15, 20.
Genet, R. M., and Hayes, D. S. Robotic Observatories, AutoScope Corp., Mesa, Arisona, 1989, pp. 205-226.
Hayes, D. S., Genet, R. M., and Seeds, M. A. 1989, in Remote Access Automatic Telescopes, ed. D. S. Hayes and R. M. Genet (Mesa, AZ: Fairborn Press), p. 169.
Seeds, M. A. 1989, in Remote Access Automatic Telescopes, ed. D. S. Hayes and R. M. Genet (Mesa, AZ: Fairborn Press), p. 163.
Seeds, M. A. 1994, I.A.P.P.P. Communiction, in press.

APT OBSERVATIONS OF SMALL-AMPLITUDE RED VARIABLES: SCIENCE AND PSYCHOLOGY

JOHN R. PERCY
Erindale Campus, University of Toronto, Mississauga, Ontario, Canada L5L 1C6

WINNIE AU
Newtonbrook Secondary School, North York, Ontario, Canada M2M 1V6

ABSTRACT   Automatic Photoelectric Telescope (APT) observations of several small-amplitude red variables, have been made as part of a multi-faceted search for and study of such variables. We comment on the important roles which both APT's and human observers can play in photoelectric photometry of variable stars.

Small-amplitude red variables (SARV's) are M giants which are pulsating with small amplitudes (generally less than 2.5 magnitudes) and with periods of up to 200 days. There are 164 known and 136 suspected variables among the approximately 500 M giants in the *Yale Catalogue of Bright Stars*. For several years, one of us (JRP) has been conducting a survey of these stars to determine the status of the suspected variables (and the M giants for which no information about variability is known), as well as to confirm the "known" variables. This has been done using the 0.4 m "teaching telescope" at the University of Toronto, operated by undergraduate student observers (Percy and Fleming 1992; Percy and Shepherd 1992) and using the American Association of Variable Star Observers (AAVSO) network of photoelectric observers (Percy et al. 1994), most of whom are amateur astronomers. This network has also been carrying out a detailed study of several dozen SARV's as part of the

AAVSO photoelectric program (Landis et al. 1992); the results of this study were reported by Percy and Yu (1994).

APT OBSERVATIONS OF K GIANTS

There are also 16 "known" and 215 suspected K giants in the *Yale Catalogue of Bright Stars*. Percy (1993) recently carried out an APT survey of a sample of 49 of these stars. Only two (HR 3275 and HR 5219) were clearly variable; they appear to be RS CVn stars. A few more program and comparison stars were marginally variable, including four early M giant comparison stars which are possible SARV's.

This project was assigned to an APT, rather than to student or amateur astronomer observers, because it was suspected that almost all of the program stars would be non-variable. Humans relish both research and discovery. Non-variability is important but not very exciting. To our knowledge, APT's do not care whether program stars are variable or not.

APT OBSERVATIONS OF M GIANTS

Several suspected SARV's, which had not been assigned to student or amateur astronomer observers, were monitored for one season with the Phoenix-10 APT as described by Percy (1993). The detailed results will be published elsewhere. Briefly, the amplitudes and periods are as follows; the periods (determined by Fourier analysis) are uncertain because of the short data sets: HR 5123 (M2 III) constant; HR 5150 (M2 III), $\Delta V = 0.09$, periods 6 days (plus others); HR 5154 (M2 III), $\Delta V = 0.08$, period 17 days; HR 5215 (M2 III), $\Delta V = 0.07$, periods 15: and 107: days; HR 5266 (M3.5 III), $\Delta V = 0.1$, periods 18:, 55: and/or >100 days; HR 5299 (M4 III), $\Delta V = 0.17$, period 30 days; HR 5300 (M1.5 III), $\Delta V = 0.1$, period 19 days.

The results of these observations confirm and strengthen the conclusions drawn from our (and other) previous surveys of SARV's:

1) The incidence of variability increases with decreasing temperature.

2) The amplitude of variability <u>tends</u> to increase with decreasing temperature, though the trend is not exact.

3) The period or time scale of variability <u>tends</u> to increase with decreasing temperature, though the trend is even weaker than for the amplitude. Any trend would, however, be weakened by the multiple time scales and/or irregularity in these stars.

The APT data provided an excellent basis for a student research project for one of us (WA). The project provided experience in data management, analysis and graphing, and an introduction to sophisticated statistical concepts such as error analysis, and power spectrum analysis. There is real potential for APT data in science education. Nevertheless, it is important for students to <u>experience</u> the process of observing with a telescope and photometer, if they are to appreciate the procedures (and pitfalls) involved.

ACKNOWLEDGEMENTS

JRP acknowledges a research grant from the Natural Sciences and Engineering Research Council of Canada, and thanks Michael Seeds and the APT Service for their excellent work. Winnie Au is a participant in the University of Toronto Mentorship Program, which enables outstanding senior high school students to work on research projects with faculty members.

REFERENCES

Landis, H.J., Mattei, J.A. and Percy, J.R. 1992, *IAU IBVS*, #3739.
Percy, J.R. 1993, *PASP*, **105**, 1422.
Percy, J.R. and Fleming, D.E.B. 1992, *PASP*, **104**, 96.
Percy, J.R. and Shepherd, C. 1992, *IAU IBVS*, #3792.
Percy, J.R. et al. 1994, *PASP*, in press.

The Four College Consortium APT: The First Four Years

Robert J. Dukes, Jr., William R. Kubinec, Harold L. Nations
Physics Department, The College of Charleston, Charleston, SC 29424

Saul J. Adelman
Physics Department, The Citadel, Charleston, SC 29409

Diane Pyper Smith
Physics Department, The University of Nevada - Las Vegas, Las Vegas, NV 89154

Edward F. Guinan, George P. McCook
Astronomy Department, Villanova University, Villanova, PA 19085

Abstract. The results of four years of observations with the Four College Consortium Automatic Photoelectric Telescope are described. Among the observing programs which have been carried out are ones involving multimode Cepheids, 53 Persei stars, spotted stars, chemically peculiar stars, Be stars, solar analogs, bright supernovae, and long period variables. A number of observations have been obtained in support of simultaneous spacecraft observations and as part of global campaigns.

1 Introduction

The Four College Consortium consists of four primarily undergraduate institutions, The College of Charleston, The Citadel, the University of Nevada, Las Vegas, and Villanova University which share the use of the Four College Automatic Photometric Telescope (FCAPT), a 30" telescope on Mt. Hopkins, AZ.. The primary observing programs are chemically peculiar stars (Adelman, Smith), Solar Type Stars (Guinan, McCook), Spotted Stars (Guinan, McCook, Nations), Multimode Variables (Dukes, Kubinec), and Be Stars (Adelman, Dukes). During four years of operation we have made more than 1,000,000 observations of objects ranging from first to thirteenth magnitude. Observations have been made in support of IUE, Voyager, and ROSAT satellite observations. Others have been part of ground based joint spectroscopy/photometric programs or in support of international campaigns. These observations have resulted in more than ten papers in refereed journals with several others in preparation as well as some 60 other presentations or publications. This work has heavily involved undergraduate students at the four institutions with at least 22 different students being involved. Thirty-two of these approximately 70 publications or presentations have had student co-authors.

2. College of Charleston

2.1. 53 Persei Stars

2.1.1. 53 Persei

Dukes participated in an extensive campaign organized by L. Huang (Beijing Observatory) on the prototype of the 53 Persei class of variables. 53 Persei has been observed during the 90-91, 91-92, 92-93, and the 93-94 seasons with the APT. M. Smith (Computer Sciences Corporation) obtained Voyager ultraviolet spectrophotometer observations during January, 1991 and January, 1992. Huang presented the results of the campaign data together with the 1990-91 APT photometry at IAU Symposium #162 on "Pulsation, Rotation, and Mass Loss in Early-Type Stars" held at Beaulieu sur Mer, France in October, 1993. It was then agreed that since the APT observations covered a much longer time span than the campaign data with very little overlap the two data sets would be published separately. Dukes has completed the reduction of his photometry which consists of 419 *uvby* observations on 145 nights. He presented these results at the Winter, 1994 meeting of the American Astronomical Society and is currently combining them with the 1993-94 photometry (Dukes, Martin, and Brannon 1994). The preliminary results indicate that the pulsation of 53 Persei is much more stable than the campaign data indicated. Comparison of the two sets of results will provide valuable insight into the relative advantages of multi-longitude campaigns over short time spans versus long series of data obtained at one location.

2.1.2. 3 Vulpeculae

Dukes, Adelman, and an undergraduate College of Charleston student (Angela Kubinec) have obtained FCAPT observations of this recently discovered 53 Persei star. Kubinec presented papers on this at the 1992 College of Charleston Research Poster session and at the 1992 meeting of the South Carolina Academy of Science (Kubinec and Dukes 1992). The latter paper won the Sigma Xi award for the best physics/astronomy presentation by an undergraduate. Adelman obtained eleven spectra with the DAO 48" telescope at a dispersion of 2.4 Å per millimeter using a Reticon detector with a typical signal-to-noise ratio of 200 and 67 Å of spectrum/exposure. His preliminary abundance analysis suggests the 3 Vul is metal poor. Photometric observations of 3 Vulpeculae were continued for the 1993 and 1994 observing seasons. W. R. Kubinec will spend a sabbatical semester in the Fall of 1994 analyzing these observations as well as some of the other APT data.

2.2. Multimode Cepheids

Dukes together with a number of undergraduates have been observing a variety of multimode Cepheids. Dukes, Gabriel Drake (a political science major), and Georgia Richardson (a physics major) have been working with data obtained on the ninth magnitude multimode Cepheid BQ Serpens during the 1991 and 1992 observing seasons. Data from the 1993 observing season is currently being incorporated. The reduction procedure has been complicated by some subtle but not insignificant errors made in constructing the request files for the APT. This has necessitated a good bit of manual re-reduction of some of the data. A preliminary periodogram analysis confirms the two known periods in the star. There has been some hint of a third period but its existence is problematical. Drake presented reports on this at the 1992 College of Charleston Research Poster Session and the 1992 meeting of the South Carolina Academy of Science (Drake and Dukes 1992).

Richardson and Dukes have analyzed data on the multimode Cepheid VX Puppis and the short period Cepheid EU Tauri. The two periods known in VX Puppis were confirmed and refined. While there is some evidence of the second period once reported in EU Tauri it is most likely an alias. Richardson won one of the six Sigma Xi Undergraduate Research awards given statewide for her presentation on this project at the annual meeting of the South Carolina Academy of Science (Richardson and Dukes 1993).

Terry Biorn (a physics major) and Dukes have begun the analysis of three seasons of data on the peculiar short period Cepheid V473 Lyrae. It is clear from this data set that if a long periodicity exists it is not the approximately 1200 day period reported by Cabanala (1991a, 1991b) but rather is more likely twice this. Obviously this star requires continued monitoring. Biorn won one of the six Sigma Xi Undergraduate Research awards given statewide for her presentation on this project at the annual meeting of the South Carolina Academy of Science (Biorn and Dukes 1993).

Dukes and Rose Forsythe (a College of Charleston undergraduate) have continued the analysis of the multimode Cepheids BQ Serpens, CO Aurigae, EU Tauri, and V473 Lyrae. Unfortunately, Forstyhe was forced to leave school for medical reasons and this project has languished. Dukes has recruited Brittany Camper (a College of Charleston undergraduate) and Lars Omberg who is a South Carolina Governor's School for Science and Mathematics Summer Research Scholar to continue the analysis of these stars during the summer of 1994.

2.3. Delta Scuti Stars

Dukes' original project involved intensive observations of the Delta Scuti stars. An extensive search of the literature and consultation with other Delta Scuti experts suggested that a worthy target was the star 4 CVn. Unfortunately this project has proven to be very difficult due to several of unforeseen factors. The most significant is due to design limitations of the telescope. Considerations such as interference by the walls of the building and the possibility of collision with other APT's have led to the establishment of hour angle limits for observation of no more than 3 hours off the meridian regardless of declination. This has prevented the 10 hour continuous runs needed for adequate observation of short period multimode variables. Most continuous campaigns are interrupted to allow observations of high priority, synoptic targets of other observers. The data gathered on 4 CVn has been reduced by Florence Foran (an education undergraduate major) and presented at the 1991 College of Charleston Research Poster Session (Foran and Dukes 1991).

2.4. Spotted Stars

H. Nations (formerly of Villanova University) has joined the College of Charleston faculty and is in the process of moving some of the stars he has been observing with the Phoenix-10 APT to the Four College APT. In doing so he will be able to observe in four filters (BVRI) instead of three (*UBV*). The addition of the observations in the red part of the spectrum gives the possibility of separating the effects of starspot temperatures and star spot areas. The stars being observed include FK Comae, HD 12545 (the most heavily spotted star ever observed) and HD 8357.

3. The Citadel

3.1 Be Stars

During the first year of operation (1990-91) Adelman (1992) obtained *uvby* photometry of 8 Be stars. This demonstrated that the telescope could produce usable photometry. As the data showed that Be stars could be variable on different time scales, he decided it would be scientifically more profitable to concentrate on fewer stars and selected FY CMa and 66 Oph as the primary Be star targets for subsequent years. In addition to *uvby* photometry, he is also obtaining measurements using a wide Hα filter to study the variable Hα emission.

Adelman and Dukes participated in a campaign involving the Be stars Psi Persei and Zeta Tauri during the 1991-92 observing season. *UBVRI* observations of these stars were obtained during the intensive phase of the campaign in September with follow up observations extending for the next several months.

In September 1992, Adelman participated in an international campaign lead by Lin Huang of Beijing Astronomical Observatory of the Chinese Academy of Sciences concerning the Be stars o And and EW Lac. Approximately 72 observations in b and y of each star were obtained over several nights. This data is being combined with photometric and spectroscopic observations taken at other observatories in several countries.

Adelman participated in the February, 1994 International Be star campaign. He observed Theta CrB as a follow-up activity through April 1994. Dr. Juan Fabregat, who is coordinating the photometric reductions, has told him that this data will be especially useful in interpreting Theta CrB's activity or lack of same during February.

3.2 HR 1105

Adelman observed the S star HR 1105 to study possible variability of a peculiar late type star. Differential photometry was obtained with Johnson B and V and Cousins R and I filters. Six observations were obtained in the fall 1991/winter 1992 observing season and 20 observations in the fall 1992/winter 1993 observing season. The star is definitely variable in B with a standard deviation of the mean variable-comparison values being 0.039 mag. compared to 0.013 mag. for the mean check-comparison values. For V the comparable values are 0.032 mag. and 0.010 mag, respectively, for R 0.026 mag and 0.009 mag, respectively, and for I 0.014 mag and 0.008 mag, respectively. This indicates that the variability decreases towards the red where the star is brighter. No period was found to have a power S/N greater than 1%. From inspection of the data, a period of order 40 days may be possible. Further observations were made during 1993/94 to try to find any periodic variability. Adelman is collaborating with Dr. H. R. Johnson, Indiana University, on the interpretation of this data.

3.3. Chemically Peculiar Stars

Adelman and Robert Fried (Braeside Observatory) found that a comparison of light curves in the $uvby$ and UBV systems showed the magnetic CP (chemically peculiar) star 56 Arietis had changed the shape of its light curves relative to those in the literature (Adelman and Fried 1993). Further there was some evidence of slight changes between values from the fall of 1990 relative to those from the fall of 1991. Additional $uvby$ observations made with the Four College APT and UBV observations with the 16-inch telescope of Braeside Observatory during the fall of 1992 and of 1993 are now being reduced. Each fall a new series of observations will be made as the probable precessional period is of order 10 years.

uvby photometry of four suspected variables, the mercury-manganese stars 53 Tauri and HR 4072, the metallic lined star, 68 Tauri, and the non-magnetic CP star HR 6096 indicated that these stars are non-variable during one or two observing seasons to within the accuracy of the observations (Adelman 1993). This data was taken during the first two years of regular telescope observation. Several papers have claimed that the first three of these stars are low amplitude variables. The use of a second comparison star, in differential photometry terminology, the check star, provides via check-comparison star measurements a guide to the accuracy of the observations which is critical for the detection of low amplitude variability. Additional observations of these stars have since been made to set more stringent limits on their variability.

During the 1993-94 academic year a senior physics major, Jack R. Knox, worked with Adelman in studying four magnetic CP stars (Adelman and Knox 1994). They refined the period of 63 And to 4.1890 days compared with 4.1920 days found by Adelman, Dukes, and Pyper (1992). The period of the sharp-lined Si star, HD 192913 is 16.840 days, a value close to, but slightly different from published values. HR 8240, a sharp-lined Si star is found not to be a photometric variable. Finally, comparison of the four color photometry of 108 Aqr with that by Morrison and Wolff (1971)using the ephemeris of North, Brown, and Landstreet (1992) shows subtle changes in the shapes of the light curves. Like 56 Ari, 108 Aqr may be precessing.

A referee misguidedly remarked that it would be scientifically more interesting to find that a HgMn star was a photometric variable than to confirm or to demonstrate that a suspected variable showed constant light. The example most often cited for variability among the HgMn stars is α And. An analysis by Adelman and his wife C. J. Adelman of FCAPT *uvby* observations definitely show it is not a low amplitude variable which is in accord with the theoretical expectation for the HgMn stars to be constant light stars (Adelman, et. al. in press).

Adelman has also been monitoring the cool magnetic CP star HR 8216 (= HD 204411) in part as neither light nor spectral variability has been discovered. It is also being studied by Miss Hulya Caliskan, Istanbul University, for her Ph.D. thesis using 2.4 Å/mm spectrograms Adelman obtained with a Reticon at the Dominion Astrophysical Observatory. Observations for three seasons indicate that this star is non-variable.

Adelman worked with two Citadel physics majors, Benjamin Brown and David Reese, in analyzing the magnetic CP stars HD 184905 and HR 8434 with FCAPT and published data. They used FCAPT observations of HD 184905 to refine the zero epoch and the period which was found to be 1.85435 days. Their photometry and that of Morrison and Wolff are generally quite similar. They found evidence for two sub-minima within the broad minimum in both u and y. HR 8434 has a period of 1.43242 days and shows a generally in phase variation of u, v, b, and y. The light curves have two nearly equally maximum and a sharp minimum. The largest amplitude is for u, 0.085 mag. (Adelman, et. al. in press).

Dr. David Bohlender, CFHT, and Adelman are working on the magnetic CP star HD 35502. A periodogram analysis of $uvby$ photometry indicates a period of about 0.85 days while the Hα emission having a period twice that length. Observations taken more frequently than once a night should be able to reconcile this difference. Adelman also has sufficiently large sets of $uvby$ data to determine periods of several other CP stars. In some cases these will be combined with those of other photometrists.

3.4. Normal Stars

Dr. Michael Joiner, Brigham Young University, asked Adelman to obtain differential photometry of 109 Vir and Theta Vir. The possible variability of these standard stars has been a concern of Dr. Benjamin Taylor for many years. Analysis from previous years suggested that these two stars were constant, but that one of the check stars was variable. This year Adelman obtained additional observations to check this possibility and also obtained multiple observations per night to check on shorter term variability.

4. The University of Nevada, Las Vegas

Pyper-Smith has been observing CU Vir which is the magnetic CP star (period = 0.5206800 days) with the shortest period known. It may experience free body precession due to its strong atmospheric magnetic field distorting the shape of the star. The precession period is many times that of the rotation period of the star, of order 5 to 10 years similar to that of 56 Ari. Three years of *UBV* data for CU Vir, obtained with the Phoenix 10-inch APT (P10) were studied and compared with previously published *UBV* data. There is some evidence for changes in shape between the P10 and earlier data. There also appear to be small differences from year to year in the P10 data, the principal of which are a slight change in the time of minimum for the 1989 data with respect to the 1987 and 1988 data. There also appear to be systematic changes in average brightness and perhaps amplitude for the *U* data from year to year. Higher precision is desirable to further pursue such studies. Some APT *uvby* data are available for CU Vir in 1991-92, but are insufficient to come to any conclusions about yearly changes in the light curves; the star is on the program for the next few years including some long nightly runs to check on cycle-to-cycle variations. Additionally, there is difficulty in fitting all the available data for CU Vir to a single period; a possible systematic change in period is being investigated. Pyper-Smith has been assisted in this by a number of undergraduates. In Spring 1992 C. DeFontaine, D. Hogge and T. Miller began reductions of CU Vir and attempted to determine the period. In summer, 1992 two REU students (UNLV was a REU site), D. Hogge and S. Young, assisted in data reduction and period determination of CU Vir using FCAPT data as well as comparing their results with previously published observations.

Period improvements have been made for a number of CP stars, for example, HD 224801 (CG And) for which there are 45 good measurements and whose period has been improved to 3.739833 +/- 0.000013 days . More than 40 observations have been accumulated for each of 8 stars in clusters and associations and analysis of these data is in progress. Pyper-Smith is also monitoring very long period CP variables (2 to 70+ years). The best example of this group of stars is HD 9996 (GY And) which undergoes spectrum and photometric variations with a period of about 8000 days. Good differential V photometry for this star is available through only one light maximum. The latest observations are in the broad light minimum; they generally agree with earlier published observations in the previous light minimum. Monitoring of this star will continue at least through the next light maximum, which, if the 8000 day period is correct, should occur in the year 2005! There is a very real question as to whether the variations of especially the very long period stars can be represented by oblique rotator models, so continuing observations of this and other very long period stars is important.

Pyper-Smith in collaboration with C. T. Bolton, P. Harmanec, R. W. Lyons, and A. P. Odell has been observing the star HD 37017 (=V1046 Ori) with the 0.8 m APT. This star is a double-lined spectroscopic binary with a B2e He-strong magnetic primary. An analysis of their results is in preparation.

Pyper-Smith has been working with David Hogge (an UNLV undergraduate) in a study of the rotation of chemically peculiar stars. and has had her Observational Astronomy class use APT data to study the W UMa type eclipsing binary, AW UMa. She has also been observing Stromgren standards with the FCAPT with the aim of improving Stromgren magnitudes and colors for comparison stars. Preliminary extinction coefficients and transformation coefficients have been determined for approximately one-third of the nights in 1992/93 and one-quarter of the nights in 1993/94 so far.

5. Villanova University

Villanova faculty currently using the FCAPT for research include George McCook, Edward Guinan, Frank Maloney, and Javad Siah. Villanova undergraduates who have participated in APT related research during this time period include Troy Thrash, Bryan Deeney, Keith Miller, David Steelman, Brian Pomerance and James Marshall.

Eight major projects are underway, many of them coordinated with satellite and other optical observing programs.

51 The Sun in Time: APT Photometry of Solar-type Stars

Guinan, McCook, and David Dorren (U. of Pa.) are continuing photoelectric photometry of several solar-type stars of different ages with the FCAPT. Several single solar-type stars are being observed to determine the evolution of stellar photospheres and to investigate activity cycles and differential rotation. Most of the program stars are being monitored in the ultraviolet with IUE and also by Sallie Baliunas (CfA) as part of the Mt. Wilson CaII H-K program. The following stars are on the Villanova *Sun in Time* photometry program:

HD 129333 (G0 V; t ≈ 70 Myr), χ^1 Ori (G0V; t ≈300 Myr), π_1 UMa (G1.5 V; t ≈ 300 Myr), HD 1835 (= 9 Ceti = BE Ceti; G2 V; t ≈ 600 Myr), κ^1 Ceti (G5 V; t ≈ 600 Myr), HD 134319 (G5 V; t ≈ 600 Myr), and HD 152391 (G8 V; t ≈ 7-8 Gyr). The presence of (photospheric) starspots in inferred from *UBVRI* or *uvby* photometry obtained in collaboration with McCook. The younger, more active stars typically show low amplitude light variations that are consistent with rotational modulation by cool starspots

All of the listed stars, except HD 152391, are younger and more active than the Sun. HD 152319 is an unusual star because it is a high velocity star and, therefore, probably much older than the Sun. However, it appears to have a rotation period of $P_{rot} \cong 11-12$ days which is about that of the Sun. Low amplitude (≈ 0.02 mag at V) periodic light variations are observed most of the time, indicating the presence of starspots on its surface. The relatively fast rotation and the presence of starspots for a single star of its assumed old age are puzzling. HD 152391 might be a binary (with an 11 day orbital period) or perhaps a former binary system that coalesced some time ago -- i.e. an old FK Comae-type star.

The youngest star in this sample, HD 129333 (= EK Dra), serves as a proxy for the zero-age-main-sequence Sun. It is probably a member of the Pleiades moving group with an age of ≈70 Myr. Photoelectric observations show it to have low amplitude (6%) light variations that imply the presence of starspots and a rotation period of about 2.7 d. There is evidence for an ≈12 year activity cycle with light and UV emission line variations hundreds of times larger than seen in the present Sun. HD 129333 represents the first single, solar-type star for which luminosity variations show evidence for a spot cycle (Dorren and Guinan 1994).

While observing the Hyades moving group star HD 134319, its comparison star, HD 135262 (G5 V) was discovered to be variable. However, it displays relatively large light variations of 0.1-0.2 mag with a characteristic time scale of ~70 days. The cause for such light variations of this apparently ordinary single G5 V star is not understood. Ground-based and *IUE* spectroscopy of HD 135262 are planned and may help explain this star's light variability.

A sample of single, solar-type stars is being studied by J. D. Dorren and Guinan as proxies for the Sun at different ages in its main-sequence lifetime. Following contraction to the main-sequence, the early Sun's more rapid rotation and consequently stronger magnetic dynamo activity is expected to result in more vigorous magnetic activity from the photosphere, chromosphere, transition region and corona. IUE (UV) and ROSAT (x-ray) observations combined with ground based *UBVRI* or *uvby* photometry offer the possibility of obtaining a detailed picture of the evolution of solar magnetic activity in its many manifestations from an age of $\approx$ 70 Myr (ZAMS) down to $\approx$ 9 Gyr (TAMS), as the solar dynamo runs through magnetic braking (Dorren and Guinan 1994).

5.2. *Apsidal Motion Studies*

Maloney and Guinan continue their study of eclipsing binary stars with eccentric orbits. Photoelectric photometry of the 10.5 day, B5V + B6V eccentric system, DI Her, was obtained with the 0.8 meter APT during June 1993. New times of primary and secondary minimum light were obtained as well as a new, well-defined light curve.

Y Cyg has been a well-known candidate for studying stellar interiors through observations of its light curve, radial velocity curve, and apsidal motion. It is a bright (7th magnitude), massive (16.7 M_O system composed of almost identical B0 V stars. Y Cyg is of the highest importance for comparison with theoretical models because the massive component stars place it in the region where the difference between models using different opacity tables is expected to be larger. The difficulty with Y Cyg lies in its dreaded almost-integral orbital period P = $2.^d 996848$. Thus, a definitive light curve is an elusive creature. Guinan, Maloney and McCook are obtaining a new, precise light curve of Y Cyg using the 0.8m APT's in Arizona. These data will be combined with those taken previously over the past two years, and those which will be taken contemporaneously at observatories at varying longitudes. Using this data we should be able to secure a reasonably complete light curve.

V 1143 Cyg has been known for some time as a prime candidate system for the study of internal mass distributions, helium abundances, and General Relativistic (GR) apsidal motion. It is composed of two F5 V stars in an eccentric orbit (e=0.54) whose period is $7.^d 64$. The systems exhibits deep and narrow eclipses, which permit very precise eclipse timings, thus yielding an accurate determination of the observed apsidal motion of $3.26\pm0.15°/100$ yr. Subtracting the theoretical GR contribution of $1.86°/100$ yr. yields a classical contribution of $1.4°/100$ yr. , implying a k_2 = 0.0041. This in turn results in a ratio of the central density to the mean density $r_c/r = 210$, or $r_c = 170$ g/cm^3 if one uses a mean density of 0.81 g/cm^3 derived from the mass and radii determinations. Comparisons with stellar interior models is quite good. If the abundances are properly chosen with X=0.70, Y=0.28, and Z=0.02, the resulting theoretical value for $k_2 = 0.0043$ is quite close to the observed value. The helium abundance is also in good agreement with the standard solar model with Y=0.27. Thus the apsidal motion study of V 1143 Cyg provides important information, difficult to obtain in any other way, of the internal structure of young, solar-type stars. V 1143 Cyg has recently been found to be a member of the Hyades moving group. This requires the age of the system to be ~ 0.6 – 0.7 Gyr. From this, we may now compare the value for the helium abundance with stellar models in a comparatively narrow evolutionary region. Guinan, Maloney and McCook are obtaining a new APT light curve as well as additional eclipse timings, and an archival search for eclipses from ~ 1900 with the Harvard Plate Collection. These new data as well as those existing in the literature will be analyzed with the addition of this newly discovered age constraint.

5.3. Studies of Red Giants and Supergiants: Alpha Herculis, Betelgeuse, and Mira

Guinan and McCook have conducted photoelectric *UBV* photometry of the M1Iab star α Her with robotic telescopes since 1988 at Mt. Hopkins, AZ. This star shows quasi-sinusoidal light variations with characteristic time scales of 75-90 days and light amplitudes of 0.2-0.5 mag with the *V* filter. This study is being coordinated with ground spectroscopy and *IUE* UV spectrophotometry being done by Myron Smith (CSC). One of the most interesting (and puzzling) results of the study is that the radial velocity variations indicate a period about 1 year, while the brightness of the star varied on a much shorter time scale of 75-95 days during 1991-92 (Taylor, et. al. 1993)

Photoelectric photometry of the prototypical M2-M7 III star, Mira, is being carried out by Guinan and McCook with automatic telescopes at Mt. Hopkins, AZ. This work is being done in collaboration with *IUE*, ROSAT, and speckle interferometry observations being conducted by Margarita Karovska (CfA). The *UBV* light curves obtained over the last five years define the long term light variability of the star, which has a mean pulsation period of ≅ 331 days. An interesting enhancement of brightness in the *U*-bandpass relative to the *B* and *V* bandpasses is apparent in the light curves. This *UV* enhancement occurs about halfway down the descending branch of the light curve and could be due to a shock wave traveling through the star.

Pointed ROSAT observations show that Mira B is a weak, but hard X-ray source. The X-rays most likely arise from accretion processes taking place as the binary companion to the M giant interacts with, and accretes from, the wind of the cool giant. The companion star appears to be a low mass main sequence star.

5.4. Sudden Dimming of FG Sge

UBVRI photoelectric photometry of the unusual variable star, FG Sge, has been carried out with the FCAPT since May 1991 by Guinan, McCook and student Troy Thrash. The 1992 observations showed quasi-periodic light variations with a period of $\cong 110^d$. The mean brightness of FG Sge during April-July 1992 was $m_v \cong +9.1$ mag with a light amplitude of 0.2-0.3 mag. Starting in mid-August, however, FG Sge began to dim rapidly. By October 1992, the visual magnitude was fainter than 13th mag. There is a strong wavelength dependence in the light loss with the largest change in brightness occurring at *V*-wavelengths and the smallest change in *U*-bandpass. Such rapid light loss has not been reported previously in nearly a century of photometric observations (Guinan, et. al. 1992)

Since that time, the FG Sge has slowly increased in brightness reaching $m_V \approx +11.0$ mag by September, 1993. In addition to the photometry, coordinated IUE observations were carried out by Guinan and collaborators David Bradstreet, Terry Teays (IUE/NASA), Benjamin Montesinos (INTA), and Charo Gonzalez (ESA). Analysis of the photometry and spectroscopy indicates that the large decrease in brightness was produced by an ejection of a dust cloud by the star similar to the dimming events observed in R CrB stars. As suggested by Joanna Jurcsik of Konkoly Observatory, FG Sge may have actually evolved into an R CrB variable.

5.5. Pre-Main Sequence Stars: V1331 Cyg, GW Ori, V773 Tau

Multi-band photoelectric observations of three pre-main sequence stars (V1331 Cyg, GW Ori, V773 Tau) are being conducted with the 0.8m APT. Guinan is studying the FU Ori-type variable, V1331 Cyg, to investigate possible outbursts and to search for periodic variability due to the rotational modulation by starspots. Villanova senior Deeney and Guinan have been investigating the long-term and short-term behavior of the weak-lined T Tauri star, V773 Tau (= HD283447).

Guinan and McCook are studying the T Tauri star, GW Ori (a spectroscopic binary, spectral type K3Ve) with an orbital period of 242 days in collaboration with Bob Mathieu (U. of Wisconsin). Recently, Shevchenko et. al. (1992) found evidence that GW Ori may be an eclipsing binary. Observations obtained so far indicate low amplitude (0.03-0.05 mag) quasi-sinusoidal light variations with a period of a few days. Photometry of GW Ori is planned over the next two years to investigate possible eclipses.

5.6. Activity Cycles in Stars with Highly Active Chromospheres

IUE observations of representative stars with highly active chromospheres are being conducted by Guinan and Dorren. These observations are being coordinated with ground-based photometry carried out with the FCAPT and with the 38-cm telescope at Villanova (student observers Deeney and Miller). IUE observations of the RS CVn binaries λ And, UX Ari, II Peg and V711 Tau (= HR 1099) were obtained during 1991/92. The UV data and ground-based photometric observations have been combined with previous observations (going back to 1978 in most cases) to investigate the long-term relationship between active regions and starspots.

The *IUE* spectra (current and archival) have been measured by astronomy students Brian Pomerance and Keith Miller. All of the program stars show evidence of systematic (possibly cyclic) variations in their chromospheric and transition-region line emissions as well as long-term variations in their respective light curves.

The long-term study of the best observed star in this sample, V711 Tau, indicates a well-defined $\cong$ 14 year periodicity in its *V*-magnitude brightness. The mean-seasonal *V*-magnitudes have an amplitude of $\approx$ 0.11 mag and had minima during 1978 and 1991/92; the star was brightest during 1984. *IUE* observations of V711 Tau during this interval are not as plentiful as the photometric observations but are sufficient to show probably cyclic variations in its chromospheric and transition region *UV* line emissions. For example, the C IV l1550 line emission is highest during 1982-83 and lowest during 1978 and 1992. The relation between the variations in the optical brightness and the chromospheric and transition-region *UV* line emissions is currently being studied by Dorren and Guinan (Dorren, et. al. 1992).

McCook and Guinan have conducted *uvby* photometry of UX Ari with the 0.8 m APT during fall 1992. These observations were coordinated with VLA C-band observations of this star carried out by N. M. Elise (USNO) and associates. Preliminary analysis of the data suggests a correlated phase modulation of the radio flux with the variations in the optical brightness of the system. In addition, McCook and Guinan obtained coordinated photometric (*uvby*) measures of V711 Tau (= HR 1099) when that star was being observed with the EUVE satellite by J. Drake (UC-Berkeley) and collaborators (Drake, et. al. 1994).

UBVRI light curves of the near-contact eclipsing binary RZ Dra were obtained by McCook with the 0.8 m APT. The light curves are being analyzed in collaboration with J. S. Shaw (U. of GA) to determine the orbital and physical properties of the binary.

5.7. HD 229041: A New Type of Stellar X-ray Source?

The 9th mag A7 III star, HD 229041, has been tentatively identified with a ROSAT PSPC X-ray source serendipitously found by Mike Corcoran (GSFC/NASA) and Javad Siah (Villanova-Physics). If HD 229041 is indeed the X-ray source, it is very unusual since at luminosity class III, X-ray emission is not known from stars of spectral types A0-G0.

HD 229041 has been placed on the FCAPT observing program by Guinan and Siah to investigate its light variability. *UBV* photometry obtained during June and September 1992 show that HD 229041 is a low amplitude ($\approx 0.^{m}06$ at *V*) variable star. This photometry, furthermore, indicates a possible ≈ 17 day periodicity in the light variation. If this periodicity is confirmed, the X-ray emission may be explained as arising from a binary companion to the more luminous A7 giant. Either a hot white dwarf or a chromospherically active cool star companion could be the X-ray source. Another possibility is that the A7 III star may *not* be an evolved star but a PMS object and, thus, the X-ray source itself. (Corcoran, Siah and Guinan 1992)

5.8. Supernova 1993J

McCook, Guinan and student James Marshall began observing Supernova 1993J in NGC3031 with the FCAPT on Mt. Hopkins in April 1993 and are obtaining UBVRI photometry data regularly. Initial publication of their findings were published in an IAU Circular (Guinan, Marshall, and McCook 1993). These observations were combined with all available photometric observations to construct a detailed light curve of the supernova during its outburst. The results of this study were presented at the 183rd meeting of the AAS in Washington, D.C. (Guinan, Marshall, and McCook 1993).

5.9. 9 Aur

During 1992/93, Guinan and McCook collaborated with K. Krisciunas (JAC, Hilo) to continue studying the light variations of the F0 V star, 9 Aur, with the 0.8 m APT. This star exhibits puzzling irregular variability in brightness of about 0.1 mag at optical wavelengths. Analysis of the photometry by Krisciunas shows the presence of periods of 1.277 d, 2.725 d, and 0.349 d. The most likely explanation of this light variation is non-radial pulsation.

6. Operation

6.1 Telescope Performance

During the 1993-94 observing season we found a higher than normal percentage of non-photometric nights. This has also been the experience of most Arizona observatories and has been attributed to the El Nino phenomenon. Despite this, the APT has operated for 167 nights out of a total of 245 in this observing season. During this period it observed for a total of 1455 hours, counted photons for nearly 800 hours and made over 260,000 measures.

Due to delays in completion of the upgrades mentioned below the telescope was late in coming on-line after the summer shutdown. We also experienced numerous crashes during the first few weeks of operation due to a subtle error in the rewritten software. We have isolated the nature of this error and have developed procedures to prevent telescope crashes until it can be eliminated. During the remainder of the 1993-94 observing season (to date) the telescope has been working well. Since observations are expected to continue for three more months final statistics for the seasons operation have not yet been compiled.

6.2. Telescope upgrades

Extensive modifications and upgrades were made to the Four College APT during the summer of 1993. These included:
1. Putting the CCD camera in a stainless steel housing with thermal insulation.
2. Adding a thermoelectric cooler to the CCD camera.
3. Providing water and glycol cooling to the hot side of the thermoelectric coolers for both the CCD camera and the GaAs photomultiplier.
4. Providing dry air (-40 degree dewpoint) flow to both the CCD camera and PMT housings.
5. Replacing the frame grabber for the CCD acquisition system with a commercial model.
6. Changing from X16 microstepping to X64 microstepping on both R.A. and Dec. axis. This was applied about half to improving the slew rate and half to improving the pointing resolution.
7. Making software changes to improve the speed of acquisition and to allow acquisition of fainter objects.

Acknowledgments

One of us (RJD) would like to express thanks to Georgia Richardson, Rose Forsythe, Lars Omberg, and Brittany Camper with help in data reduction. All of us would like to express our heartfelt appreciation to Lou Boyd for keeping the telescope operating and to Russ Genet for getting us involved to start with. This research is sponsored in part under the following grants: NSF AST-8616362, NSF AST-9115114, USE-915614 and NASA NAG 5-382.

References

Adelman, S. J. 1992, *Publ. Astr. Soc. Pacific*, **104**, 392

Adelman, S. J. 1993, *Astron. Astrophys*, **259**, 411

Adelman, S. J., R. J. Dukes Jr. and D. M. Pyper 1992, *Astron. J.* **104**, 314

Adelman, S. J. and R. Fried 1993, *Astron. J.* **105**,1103

Adelman, S., B. H. Brown, H. Caliskan, D. Reese and C. J. Adelman in press, *Astronomy and Astrophysics*

Adelman, S. J. and J. Knox J. R. 1994, *Astronomy and Astrophysics*, **103**, 1

Biorn, T. and R. J. Dukes Jr. 1993, *Bulletin of the South Carolina Academy of Science*, **LV**, 65

Cabanela, J. E. 1991, *J. AAVSO*, **20**, 54

Cabanela, J. E. 1991, *Robotic Observatories: Present and Future* (Mesa, AZ: Fairborn Press), 317

Corcoran, M. F., J. Siah and E. F. Guinan 1992, *B.A.A.S.*, **24**, 1152

Dorren, J. D. and E. F. Guinan 1994, Ap J, **428**, 805

Dorren, J. D. and E. F. Guinan 1994, *The Sun as a Variable Star* (Cambridge: Cambridge Univ. Press), 207

Drake, G. and R. J. Dukes Jr. 1992, *Abstracts of the College of Charleston Scientific Research Poster Session,*, **4**, 29

Drake, J. J., A. Brown, R. J. Patterer, P. W. Vedder, S. Bowyer, G. E. G. and R. F. Malina 1994, *ApJ Letters* **421**, L43

Dukes, R. J., C. Brannon and R. Martin 1994, *B. A. A. S.* **25**

Foran, F. and R. J. Dukes Jr. 1991, *Abstracts of the College of Charleston Scientific Research Poster Session,*, **3**, 17

Guinan, E. F., M. J. and G. P. McCook 1993, *IAU Circ.* 5770

Guinan, E. F., J. Marshall and G. P. McCook 1993, *B. A. A. S.* **25**, 1349

Guinan, E. F., G. P. McCook and T. A. Thrash 1992, *IAU Circ.* 5632

Kubinec, A. J. and R. J. Dukes Jr. 1992, *Bulletin of the South Carolina Academy of Science* , **LIV**, 83

Morrison, N. D. and S. C. Wolff 1971, *Publications of the Astronomical Society of the Pacific* , **83**, 474

North, P., D. N. Brown and J. D. Landstreet 1992, *Astronomy and Astrophysics*, **258**, 389

Richardson, G. A. and R. J. Dukes Jr. 1993, *Bulletin of the South Carolina Academy of Science*, **LV,** 110

Shevchenko, V. S., K. R. Grankin, M. A. Ibragimov and S. Y. Melnikov 1992, *IBVS*, 3746

The Fairborn/TSU Robotic Telescope Operations Model

Gregory W. Henry

Center of Excellence in Information Systems, Tennessee State University, 330 10th. Avenue North, Nashville, TN 37203

Abstract. The operation of multiple robotic telescopes at Tennessee State University is described. These telescopes are all located at a single site, and each telescope is dedicated to a single, long-term observing program. These features result in an automated data acquisition, reduction, and archiving system that is extremely productive, yet inexpensive to operate and easy to manage.

1. Introduction

For the past several years, Tennessee State University has operated three telescopes for automated photoelectric photometry at Fairborn Observatory's APT site in southern Arizona (Genet et al. 1987; Genet, Boyd, & Baliunas 1986). Fairborn Observatory is a non-profit organization headed by Lou Boyd for the development and operation of robotic telescopes. Its observing site is located at the Smithsonian Institution's Fred L. Whipple Observatory on Mt. Hopkins in the roll-off-roof enclosure of the Smithsonian's old satellite tracking installation at 7800 feet above sea level (Burke & Kirchhoff 1968). The first of the three telescopes to go into operation (in early 1986) was the Fairborn 10-inch. By the end of 1994, this telescope had collected over 49,000 group observations of variable stars, the majority of which are semi-regular variables in a collaborative program with the Harvard-Smithsonian Center for Astrophysics. These group observations include all comparison star, check star, sky, and dark integrations in all desired filters needed to determine differential magnitudes for a given program and check star. The second telescope, a 16-inch, began operating in November 1987 and has collected 72,000 group observations of chromospherically active single and binary stars in a collaboration with Vanderbilt University. The third is a 30-inch telescope that has made 11,000 high-precision observations of solar-type stars since April 1993, also in collaboration with the Center for Astrophysics. These three telescopes and the precision of the data acquired with them are described in my paper "The Development of Precision Robotic Photometry", also in this volume. A fourth telescope, a 32-inch dedicated to high-precision photometry of solar duplicate stars, will go into operation in early 1995.

2. Description of the Operations Model

Two primary features that define the Fairborn/TSU robotic telescope operations model are (1) all telescopes are located at a single site and (2) each telescope is dedicated to a single long-term observing program. Our operation uses only one person (Boyd) at the remote site to develop, maintain, and oversee the operation of the telescopes and one person (Henry) at Tennessee State University who is responsible for scheduling the observations on the telescopes, performing daily quality control checks, reducing and archiving the data, and troubleshooting problems with telescope performance. We have designated these two positions the Principal Engineer (PE) and Principal Astronomer (PA), respectively. Scientific collaborators are also essential to help digest the tremendous flow of data that results from this operation, which we feel is the simplest and most effective way to operate small robotic telescopes.

There are several advantages to having all telescopes located at the same remote site and within a single enclosure. Physical plant requirements such as road access, power (and power backup systems), phones, computer network connections, site maintenance and security are all greatly simplified when only one site needs to be supported. A single set of weather sensors and a single site control computer for monitoring weather conditions and for opening and closing the roof can service all of the telescopes and protect them when the weather deteriorates. A single communications computer can forward observing requests from the PA to the appropriate telescopes, gather the nightly data from each telescope, and forward them to the PA each morning over the Internet. Workshops, tools, and electronic test equipment can be can be easily available on site. Instrumental support systems such as power supplies, air filtration and drying equipment, and temperature control systems can be centralized and support all telescopes. Perhaps the most significant advantage is the ability of a single PE to manage the site in support of all telescopes. In fact, this has been done by Boyd for a decade in his spare time!

Additional advantages arise when each telescope managed by the PA is dedicated to a specific, long-term observing program. Each program can benefit from large quantities of high-quality data taken year-round from an excellent observing site. Because these data are taken with dedicated instruments, they benefit from an internal consistency not possible with traditional manual observing runs on multi-user telescopes. Standardized observing sequences on each telescope that are optimized for each project make it simple to program the telescopes to make the observations but, more importantly, make data reductions straightforward and efficient since automated reduction and archiving programs can be written that require very little effort by the PA to oversee. Also, standard star and other quality control observations tailored to the specific observing program can be developed to ensure that the data are of the highest possible quality, and the reduction of these observations can also be automated. With only one scientific program running on each telescope, complete sets of standard star and quality control observations can be made each night on each telescope while sacrificing only a few percent of the observing time. With the automated reduction routines, a PA can process the nightly output of several telescopes in only a few minutes each day. Additional advantages in terms of cost and quantity of data produced are reviewed below.

There are admittedly a few disadvantages of our robotic telescope operations model, and there are several places where the operation is subject to single-point failure. Obviously, if it is cloudy over Mt. Hopkins, we will not get data from any of our telescopes. We do not benefit from a distribution of our telescopes in longitude around the earth or in complementary weather patterns at multiple sites. Power failures, ice storms, failure of the site control computer to open the roof, failure of centralized cooling systems, etc. all can similarly result in the total shutdown of our system. A couple other possible sources of single-point failure are simply too disquieting to mention. In addition, telescopes dedicated to long-term programs cannot be very responsive to short-term or target-of-opportunity projects without some sacrifice of the long-term projects. On the positive side, however, our telescopes have been up and operating for the past several years on well over 95% of the available clear nights on Mt. Hopkins, and our observational programs have benefited tremendously as a result. Tennessee State University recently made an attempt to overcome the difficulties inherent in a multi-site operation and eliminate the single-point failure modes by placing a 32-inch APT on Mt. Wilson, but the failure of AutoScope Corporation to produce a working instrument (Henry 1995) forced us to retreat to our single-site operation. For now, this seems to be the only practical option available to us.

It can also be mentioned that other robotic telescopes at the Fairborn site are operated in modes other than our own. One 30-inch APT is owned and operated by a consortium of four universities, and the observing time is shared among several astronomers, one of whom serves as PA for the telescope (Dukes et al., "The Four College Consortium APT: The First Four Years", this volume). Two additional 30-inch and another 10-inch APT are owned by Fairborn and operated in a "rent-a-star" mode whereby observing time is purchased by a large number of observers on a per-observation basis, again with a single PA for each telescope to oversee the scheduling of observations and distribution of data (Seeds 1994; Seeds, "The Phoenix 10 Automatic Photometric Telescope: An Update", this volume). Boyd also serves as PE for these telescopes.

3. The ATIS Standard Programming Language

An implicit part of the operations model described here is the language in which the individual robotic telescopes are programmed. In 1989, Fairborn Observatory developed the Automatic Telescope Instruction Set (ATIS) to define the observational requests for the Fairborn APT's (Genet & Hayes 1989). For the first time, this new language allowed APT users to tailor the exact specifics of their observing sequences, to communicate them to the telescopes via ASCII file transfers over telephone lines or the Internet, and to retrieve the resulting observations the next day. In ATIS, a group observation is the primitive unit to be scheduled and executed. The ATIS groups consist of sequences of telescope and instrument commands composed by an astronomer to accomplish a given observation and contain commands to move the telescope, to acquire and center stars, to control the filters, and to make integrations in a specified sequence.

In addition to specifying the syntax and semantics for observation requests and results, the ATIS language provides a set of group selection rules that are used to determine the execution order of groups during the night. The group selection rules provided by ATIS essentially implement a first-to-set-in-the-west policy: at any given point in time the telescope observes the star that will next move out of the hour angle limits in which it can be observed. Through the use of user-input parameters to define group types, group priorities, probabilities of execution, date and local sidereal time limits, number of observations requested, etc., ATIS allows APT users to define any desired observation sequence as well as to intersperse standard star observations and other quality control checks throughout the night and have them all scheduled automatically. These group selection rules allow our robotic telescopes to run for long periods of time, automatically selecting targets to observe each night in the face of changing seasonal availability of stars and interruptions due to bad weather.

However, in spite of the extraordinary improvements to the long-term scientific programs being conducted at Tennessee State University and elsewhere that have resulted from the use of automatic telescopes running the ATIS control language, the 1989 version of ATIS is not without its limitations. For instance, there is no look-back or look-ahead capability that might help the telescope decide when to make a given observation, perhaps to fill in a variable star's light curve automatically or make an observation that got missed on a previous night. Many situations can arise, particularly on multi-user telescopes, when more intelligent scheduling than the simple ATIS group selection rules can provide is needed so that the highest priority observations are made at scientifically appropriate times and all users are treated fairly in the allocation of telescope time. Also, since ATIS only allows access to the telescope before the beginning of the night and after the end of the night, no information is available during the night about the current status of the observing program, the quality of the data being obtained, or about the performance of the telescope and detector that might allow appropriate modifications of the observing program to be made in real time.

To address these limitations of the original ATIS control language, Fairborn Observatory collaborated with a committee of APT users and with the artificial intelligence (AI) group at the NASA Ames Research Center to develop a new version of ATIS with the capability to override default ATIS group selection rules. A mechanism was developed for communication with a telescope during the night using incremental ATIS partial input and partial output files. This new feature makes it possible to implement an external AI scheduler (running on a remote machine) that can effectively drive the telescope controller in near real time to improve the scheduling of the observations. The result, including additional enhancements to ATIS, was published as ATIS93 by Boyd et al. (1993) in hopes of its becoming an international standard language for automatic telescope control. The paper by Drummond et al. ("Flexible Scheduling of Automatic Telescopes over the Internet", this volume) describes how ATIS93 will be incorporated into a new set of software tools called the Associate Principal Astronomer (APA) being developed at NASA Ames that will include an AI scheduler and greatly facilitate the management of robotic telescopes.

4. Comparison with Old Operations Model

It is interesting to compare our robotic telescope operations model with previous manual observations in terms of the level-of-effort and costs per observation. For the manual observing discussion, I examined old observing logs from 1980, a year when I was fully employed by Vanderbilt University and my first priority was to make manual photometric observations of chromospherically active stars. During that year I had essentially unrestricted use of the Dyer Observatory 24-inch telescope to use whenever it was clear. Nashville is not the most promising site from which to do photometry; my observing log shows that I observed at Dyer on 40 different nights during 1980 and made a total of 396 complete differential group observations. All 40 nights were not clear all night; the number of observations per night ranged from 1 (when it became non-photometric shortly after I began) to 28 (when it was clear all night long). The average number of observations per night was 10.

Because of the relatively poor observing prospects in Nashville, I applied for two observing runs on Kitt Peak with the No. 4 16-inch telescope. Of the 42 nights I received in 1980, I was able to observe on 34 and made a total of 601 group observations. The number of observations per night ranged from 1 to 32 with an average of 18. The higher average compared to Dyer is expected since the chances for a full night of observing at Kitt Peak were higher than at Dyer. The actual efficiency of manual observing with the two telescopes was comparable; full nights of observing resulted in approximately the same number of observations from each telescope. However, it was apparent that a few weeks of observing time at a remote but good site resulted in more observations than could be acquired during the rest of the year from the home site located in less favorable observing conditions (601 vs. 396, respectively).

In sum, I observed on 74 nights in 1980 and obtained 997 group observations. I estimate that I spent approximately 50% of my time obtaining these observations. This included preparation of proposals to Kitt Peak, travel, observing, preparation of finding charts, wasted effort on nights that turned non-photometric, data reduction (which in those days was nearly as much drudgery as the actual observing), and recuperation. Therefore, roughly 1000 differential group observations per year were made and reduced by an observer working approximately half-time from a typical eastern site with occasional trips to a remote site while using typical manual telescopes and data logging systems common 15 or 20 years ago.

For comparison with automated observing, I examined a recent 12-month interval when the 10-inch, 16-inch, and 30-inch APT's were fully operational. During this time, the Fairborn 10-inch secured 7427 group observations, the Vanderbilt/Tennessee State 16-inch made 13,952 group observations, and the Smithsonian/Tennessee State 30-inch obtained 7246 observations. The total number of group observations from all three telescopes during this 12-month interval was 28,625.

However, the typical APT group observation was more complex than the typical manual group observation from 1980. In particular, more check and comparison stars were used, observations were made in more filters, and many more standard star and quality control observations were made with the APT's. Indeed, the average APT group observation represents at least twice the data as the average manual group observation. So a very conservative comparison reveals that our robotic telescopes produce fifty times more data per year than one of my best years of manual observing. This will increase even further when our 32-inch APT goes into operation in 1995.

Once the automated telescopes are built, installed, debugged, and working smoothly, and the long-term observing programs are set up to run on them, and the automated data reduction, quality contol checks, and archiving software are written, very little time is required on a daily basis to manage the telescopes. I certainly spend less than 10% of my time on this and probably considerably less than that during normal weeks. Boyd estimates that he spends at most a few percent of his time on maintenance of these three telescopes. So, again being very conservative, we can say that in one-fifth of the time spent, we make 50 times more observations with our robotic telescopes than was possible with previous manual techniques. This corresponds to a 250-fold increase in data acquired per unit time spent by the astronomer.

One caveat should be expressed here. The kind differential photometric observing I have been discussing involves hundreds of moves between stars each night with relatively short (10 - 20 sec) integration times on each star. Moving the telescope, identifying the proper star, centering the star, moving filters, starting integrations, and logging data are all operations an automated system can do much more efficiently than a manual observer. However, for observing programs involving much longer exposure times (such as high-resolution spectroscopy), a manual observer could work nearly as efficiently as an automated system, at least until he fell asleep. So not all kinds of observing will show the same gains in observing efficiency as have been achieved with automated differential photometry. Even so, there are still many good reasons for automating other kinds of observing programs (Eaton, this volume).

Finally, we can now compute a cost comparison between our automatic observing system and earlier manual observing. For the 12-month period cited here, the 16-inch APT, for instance, made 13,952 group observations. Fairborn operates and maintains this telescope for $15,000 per year. Therefore, the cost per observation was $1.07 per group. To operate a 16-inch telescope manually on every clear night of the year, (at least) two full-time observers would be needed. At $30,000 per year each plus benefits and overhead, this would cost $90,000 per year. On a typical good night, the 16-inch APT makes approximately 100 group observations. On my very best nights of manual observing, I could make only one-third that number, and each manual group observation contained only one-half the data. Therefore, the manual observers will get one-sixth the number of observations per year, compared to the APT, and at six times the cost. Therefore, the cost per observation obtained manually would be 36 times the cost of an APT observation.

All of these considerations, combined with the order of magnitude improvement in the precision of APT observations (Henry, "The Development of Precision Robotic Photometry", this volume) dramatically illustrate that the operation of automated telescopes for dedicated, long-term differential photometric observing programs (i.e. the Fairborn/TSU model) has brought revolutionary advances to this field.

Acknowledgments. The development and operation of robotic telescopes and the analysis of data from them has been supported for several years at Tennessee State University by the National Aeronautics and Space Administration and by the National Science Foundation, most recently through NASA grants NAG 8-1014 (Marshall Space Flight Center) and NCC 2-883 (Ames Research Center) and NSF grant HRD-9104484. None of this work would have been possible without the dedicated efforts of Lou Boyd at Fairborn Observatory.

References

Boyd, L., Epand, D., Bresina, J., Drummond, M., Swanson, K., Crawford, D., Genet, D., Genet, R., Henry, G., McCook, G., Neely, W., Schmidtke, P., Smith, D., & Trueblood, M. 1993, IAPPP Comm. No. 52, p. 23

Burke, J. J. & Kirchhoff, W. 1968, Sky & Tel., 36, 284

Genet, R. M., Boyd, L. J., & Baliunas, S. L. 1986, IAPPP Comm. No. 25, 15

Genet, R. M., Boyd, L. J., Kissell, K. E., Crawford, D. L., Hall, D. S., Hayes, D. S., & Baliunas, S. L. 1987, PASP, 99, 660

Genet, R. M. & Hayes, D. S. 1989, Robotic Observatories (Mesa: AutoScope)

Henry, G. W. 1995, IAPPP Comm. No. 57, p.74

Seeds, M. A. 1994, IAPPP Comm. No. 56, p. 23

The Development of Precision Robotic Photometry

Gregory W. Henry

Center of Excellence in Information Systems, Tennessee State University, 330 10th. Avenue North, Nashville, TN 37203

Abstract. A robotic telescope development program has been underway at Tennessee State University for the past several years, driven by the desire to measure smaller and smaller stellar luminosity changes. The goal of this program has been to increase the precision and efficiency of photometric observations from robotic telescopes as well as to automate the reduction and archiving of the resulting data. Through the use of larger telescopes, precision photometers, new observing strategies, and detailed quality control monitoring, we have succeeded in improving the precision of robotic telescope photometry by an order of magnitude. The internal precision of observations with our newest telescopes matches the expected photon and scintillation noise limits, and the rigorous quality control made possible by automation results in a level of external precision unmatched with traditional manual telescopes.

1. Introduction

For the past several years, observational programs at Tennessee State University in chromospherically active stars, solar-type stars, solar duplicates, and chromospheric structure have made it clear that our understanding of stellar magnetism is severely constrained by the lack of sufficient observational data. We have attempted to alleviate this situation somewhat through the development of automated photometry. We are also planning for the development of automated spectroscopy (see Eaton,"The Rationale for an Automatic Spectroscopic Telescope," this volume).

This paper describes the operation of four automatic photoelectric telescopes (APT's) in order to document the gains made in performance of these systems in recent years. All four telescopes, along with others, are located under a single roll-off-roof shelter at the Fairborn Observatory (a non-profit organization headed by Lou Boyd for the development and operation of robotic telescopes) on Mt. Hopkins (Figure 1; Eaton, Figure 1, this volume). These telescopes have brought revolutionary improvements to our long-term photometric monitoring programs (see Henry, "The Fairborn/TSU Robotic Telescope Operations Model," this volume). We have acquired over 150,000 sets of differential observations with these telescopes, and, with collaborators at various institutions, have published over 100 scientific papers over the past six years dealing with APT results.

Figure 1. The Fairborn Observatory APT site on Mt. Hopkins. On the right is the roll-off shelter housing 8 APT's. On the left is the control building containing computer systems, workshops, and other support facilities.

2. The Fairborn 10-inch APT

The Fairborn 10-inch APT (Figure 2) was the first of (currently) eight APT's to go into operation on Mt. Wilson. The tube assembly and optics are from Meade Instruments, the mount is from DFM Engineering, and the SSP-3 photometer (with an ambient temperature photodiode detector) is from Optec. Since early 1986, this telescope has collected over 49,000 group observations of variable stars through Johnson V, R, and I filters in a joint observing program with the Harvard-Smithsonian Center for Astrophysics. The observations are conducted in a fixed group sequence: K,S,C,V,C,V,C,V,C,S,K where K is a check star, S is a sky position, C is a comparison star, and V is the variable star. We are restricted to this fixed sequence by the telescope's primitive (compared to more recent APT's) control system that cannot be programmed to handle other kinds of observing sequences.

The precision of the data acquired with this telescope is limited by several factors. (1) The small aperture of the telescope causes significant scintillation noise. Young (1974) provides a method of estimating the scintillation noise of an observation as a function of telescope aperture, observatory altitude, integration time, and air mass. In the case of the 10-inch telescope on Mt. Hopkins (7800 ft. above sea level), the scintillation noise for a group mean taken at a median airmass of 1.35 with our 10 second integration times would be about 0.0035 mag, after correction for a factor of two error in Young's original formula

Figure 2. The Fairborn 10-inch APT. This telescope is being used by Tennessee State University and the Harvard-Smithsonian Center for Astrophysics to monitor a variety of variable stars, especially semi-regular variables.

(Young 1991). (2) The *internal precision* (i.e. the repeatability of successive differential magnitude measurements within a group) is also limited by imperfections in the centering process. The telescope centers stars in the focal-plane diaphragm of the photometer by taking counts with the photometer in four overlapping regions of the sky centered around the presumed position of the star (Boyd, Genet, & Hall 1984). The counts in these four regions are compared, the direction to the true star position derived, and the telescope moved in that direction. This process is repeated until the count rates in all four regions are the same, indicating that the telescope has properly centered the star. The centering accuracy is only one-fourth to one-third of the diaphragm diameter. (3) Wear in the telescope's worm-gear drive systems resulting from heavy use in the automated observing mode places additional limits on the internal precision. This wear makes it difficult for the preloaded drives to resist motion due to wind loading or slight imperfections in telescope balance and so aggravates the already marginal centering accuracy. (4) The primitive control system precludes the possibility of measuring standard stars to determine nightly extinction coefficients. Therefore, the *external precision* (i.e. the repeatability of night-to-night group mean differential magnitudes of constant star pairs) is degraded because the data must be reduced with long-term mean extinction coefficients determined on occasional nights of standard star observations. (5) The photometer undergoes seasonal and night-to-night changes in color sensitivity due to ambient temperature changes. Again, because nightly standard stars cannot be

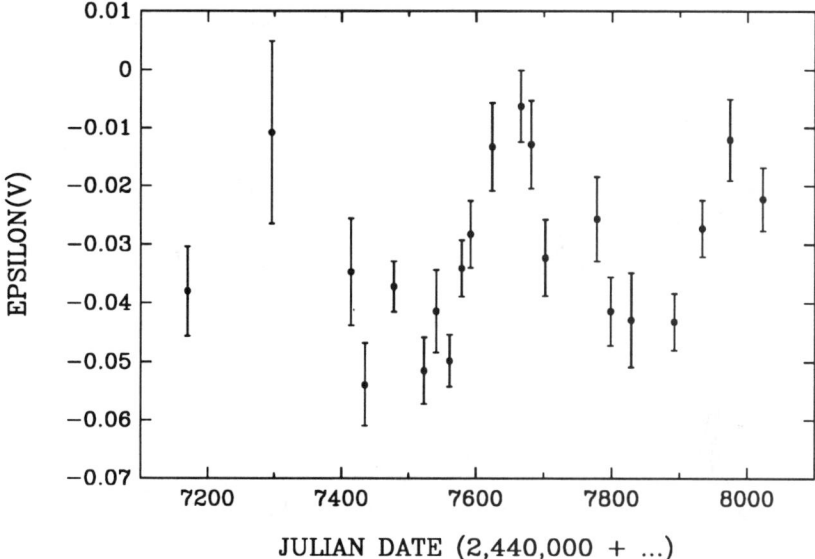

Figure 3. Johnson V transformation coefficients for the Fairborn 10-inch APT. Seasonal temperature variations cause significant variations in color response of the ambient temperature solid-state detector.

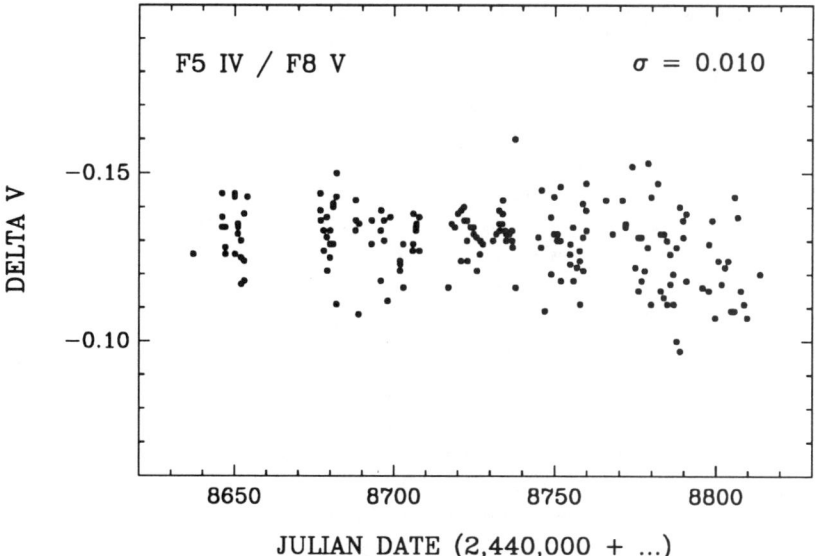

Figure 4. Nightly group mean differential magnitudes in V for the pair of constant stars HD 125451/HD 124570. The standard deviation of the nightly observations from the seasonal mean provides a measurement of the external precision of the Fairborn 10-inch APT.

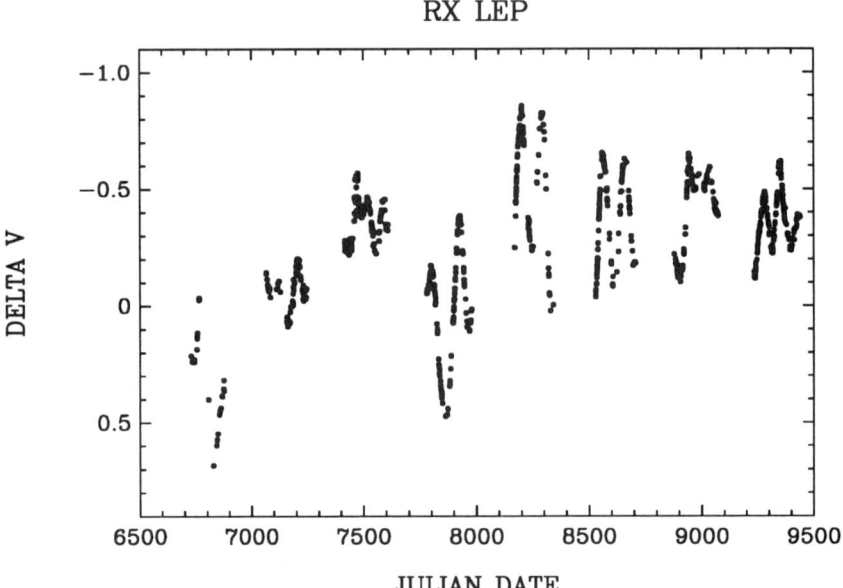

Figure 5. Eight year light curve of the semi-regular variable RX Leporis. Periodogram analysis finds 58 and 89-day pulsation timescales in the data.

observed, the data are reduced with mean transformation coefficients. Figure 3 shows the effect of seasonal temperature variations on Johnson V transformation coefficients determined occasionally over a two and one-half year period.

Because the 10-inch APT (as well as our other APT's) has no way of knowing the quality of a night in real time, it collects data as long as it can find stars. Therefore, a way must be found to eliminate *ex post facto* data that were taken in non-photometric conditions. The stars we typically observe with APT's are not expected to vary significantly over the few minutes required to obtain a group observation. Therefore, we use the internal precision of a group observation as measured by the standard deviation of the group mean differential magnitudes as our cloud filter. In good photometric conditions, these group mean standard deviations average about 0.005 to 0.007 mag. Thus, to filter the data at the 3-sigma level, we eliminate any group mean with a standard deviation larger than 0.02 mag. This cloud filter works extremely well and few errant points survive the filtering process. The fact that our internal errors are considerably larger than the scintillation noise estimated above, even for bright stars where photon noise is 0.001 mag or less, indicates that a substantial fraction of these errors is due to coarse centering of stars in the diaphragm.

To determine the external precision of data from the Fairborn-10, we measure pairs of constant stars each night. Figure 4 displays data on the constant pair HD 125451 / HD 124570 from one observing season. Each point in the figure represents one nightly group mean differential magnitude in the V band, and the 0.010 mag standard deviation of the nightly observations from the seasonal mean represents the external precision. Essentially the same precision is obtained in the R and I filters. This matches the photometric precision documented by Young et al. (1991) for the Phoenix-10 APT which began operations more than 10 years ago in Boyd's back yard and was subsequently moved to Mt. Hopkins in 1986.

This 1% precision is adequate to produce beautiful long-term light curves of semi-regular variable stars with amplitudes of a few tenths of a magnitude. Figure 5 shows eight seasons of V data on RX Leporis. These data and data from nine other semi-regular variables were analyzed recently by Cristina et al. (1995). Several of the stars, including RX Leporis, show two or three distinct timescales of pulsation, so this type of long-term monitoring is just beginning to reveal the true pulsational characteristics of these stars.

3. The Vanderbilt/Tennessee State University 16-inch APT

The VU/TSU 16-inch APT (Figure 6) went into operation in November 1987 to study chromospherically active stars, which vary typically by only 0.1 mag or so. DFM Engineering manufactured the mechanical components of the telescope. Paul Jones of Star Instruments produced the optics, and Lou Boyd at Fairborn designed and built the control system and photometer, which uses an EMI 9924B photomultiplier tube. This joint Vanderbilt/Tennessee State collaboration has produced over 72,000 Johnson B and V group observations of over 200 stars. Some of the early problems encountered in the operation of the 16-inch APT have been described in Hall & Henry (1993).

During its first three years, the 16-inch was operated in a mode identical to the Fairborn-10 described above. At the end of the third year, the original worm-gear drives on both axes were replaced with micro-stepping belt-drive systems designed by Boyd. These belt-drive systems provide better resistance to wind loading, are free of backlash, allow much faster slewing of the telescope, and eliminate the mechanical wear associated with the worm drives. At the time the drives were upgraded, a new control system was installed that is capable of running the Automatic Telescope Instruction Set (ATIS) (Genet & Hayes 1989) developed at Fairborn. With ATIS, we have the ability to modify our differential observing sequences and integration times as needed, although we continue to use primarily the standard sequence described above for the Fairborn-10. After four years of operation, the original photometer was replaced by a new precision photometer designed and built by Boyd. This new photometer features a CCD camera for finding and centering stars in the diaphragm. Centering times improved from around 10 to 15 seconds with the old centering routine (described above) to about two seconds with the CCD camera. Centering accuracy also improved to approximately two arcseconds. Once a star is properly centered in the diaphragm, a rotating mirror moves out of the light path to admit light to the photomultiplier. Filtered and dried air is constantly circulated inside the sealed enclosure of the photometer to control dust and humidity. Accurate temperature control is provided by a combination of thermoelectric coolers and circulating coolant that stabilize the enclosed photomultiplier, filters, and electronics to within a degree or so Fahrenheit. Two rotating filter wheels and one rotating diaphragm wheel provide a large selection of color and neutral density

Figure 6. The Vanderbilt/Tennessee State 16-inch APT. The precision photometer built by Boyd is contained in the dark tub on the tail of the telescope.

filters and diaphragm sizes. The improved efficiency of the belt drives and CCD centering allowed us to place approximately 50% more program stars on the observing menu than we were able to observe previously.

Since the precision photometer went into operation in early 1992, we have used ATIS to schedule nightly observations of standard stars selected automatically from a master list of standards covering the observable sky. A least-squares fit to each night's standard star observations produces a simultaneous solution for the nightly extinction coefficients, transformation coefficients, and zero points. These nightly coefficients are accepted and saved only if a sufficient number of standard star observations are made over sufficient ranges of airmass and color index, and if the rms of the all-sky solution is less than 0.03 mag. The differential reductions of the program star data then proceed with the nightly extinction coefficients, if available, or a running average of the three most recent nightly extinction coefficients if the standard star solution was not satisfactory.

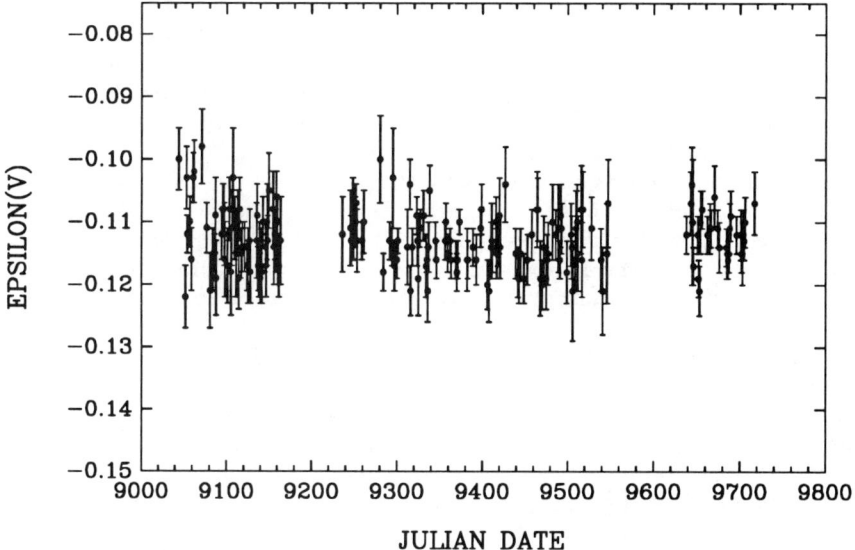

Figure 7. Nightly transformation coefficients determined by observations of standard stars with the 16-inch precision photometer. The lack of seasonal variation compared to Figure 3 shows that tempererature control of the photometer has eliminated seasonal variations in transformations.

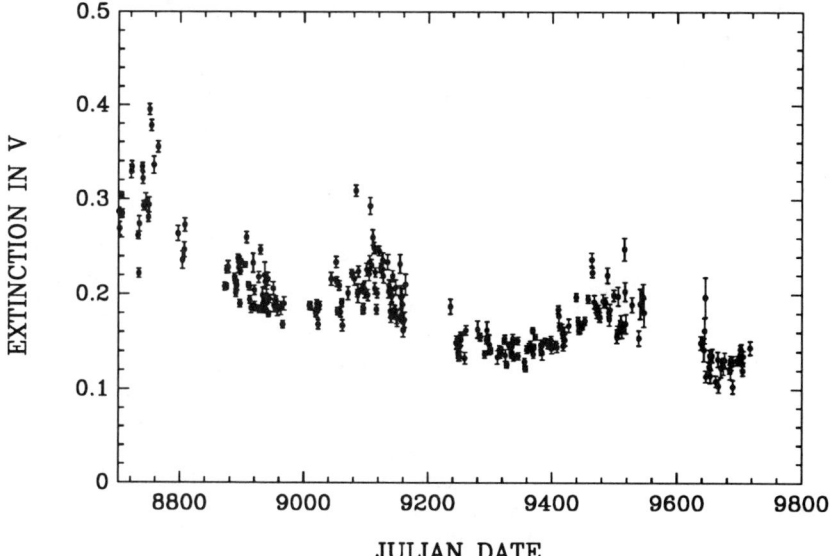

Figure 8. Nightly extinction coefficients (in mag/airmass) from the 16-inch APT. Seasonal variations of extinction are superimposed on a slow decline in extinction as the atmosphere clears after the Mt. Pinatubo eruption.

Only long-term averages of the nightly transformation coefficients are used to convert our differential magnitudes to the standard system. The internal precision, i.e. the average nightly standard error of the group means, now averages between 0.0025 and 0.0035 mag on good nights, so only observations that survive a 0.01 mag (3-sigma) cloud filter are saved. All reductions occur each morning under control of automated routines developed for a PC.

Nightly transformation coefficients for the past two years are plotted in Figure 7. Gaps in the data occur each year when the site is shut down for protection against intense lightning storms prevalent in July and August. Compared to results in Figure 3 from the Fairborn-10, it is obvious that careful temperature control of the precision photometer has eliminated seasonal variations in the transformation coefficients.

Nightly extinction coefficients in V are shown in Figure 8. This record begins approximately six months after the gigantic eruption of the Mt. Pinatubo volcano in the Philippines in June 1991. When the resulting dust cloud reached southern Arizona, extinction coefficients tripled. Since then, the usual seasonal variations in extinction have been superimposed on a gradual decline due to the dissipation of dust and high-altitude aerosols from Pinatubo. Extinction finally returned to normal levels in late 1994.

Nightly zero points (in magnitudes) are plotted in Figure 9. While not used in the differential reductions, these zero points are extremely useful for monitoring the stability of the system and for deciding when to clean the optics. A linear decrease with time over several months is generally seen as dirt accumulates on the mirrors and photometer entrance window, followed by a jump when the optics are washed.

Since the precision photometer was installed, the internal precision of the 16-inch APT typically ranges from 0.0025 to 0.0035 mag. Young's scintillation model predicts 0.0025 mag of scintillation noise for a typical group mean from the 16-inch. This, combined with a photon noise of 0.001 mag for a typical group observation, yields an expected internal precision of 0.0027 mag, in good agreement with what we actually observe. Therefore, the internal precision of our observations is no longer limited by centering or tracking problems, but can be completely accounted for by scintillation and photon noise.

An eight-year history of the external precision from the 16-inch is written in the observations of the constant red/blue pair 27/28 Leo Minoris (Figure 10). The first four seasons of data were taken with the original, ambient-temperature photometer. The external precision increased over the first three years from 0.009 mag to 0.016 mag due to wear in the worm-gear drives. Precision in the fourth season improved to 0.008 mag with the installation of the new belt-drive systems, but the original photometer and crude centering algorithm were still in use. The precision photometer was installed prior to the fifth season. The improved CCD centering and temperature stabilization of the new photometer plus the use of nightly extinction coefficients whenever available improved the external precision to between 0.003 and 0.004 mag. This is only slightly larger than the internal precision of the observations, in spite of the 1.0 mag color difference between these two stars.

Examples of observations over different timescales with this telescope are shown in Figures 11 - 13. Figure 11 displays a light curve constructed from three

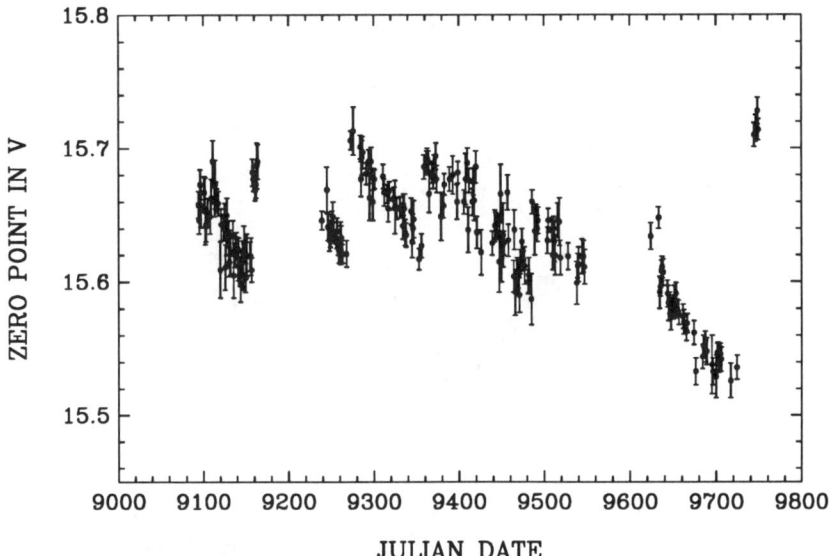

Figure 9. Zero points (in mag) from all-sky solutions of nightly standard star observations. These are useful for determining (from 1600 miles away) when it is time to have the optics cleaned.

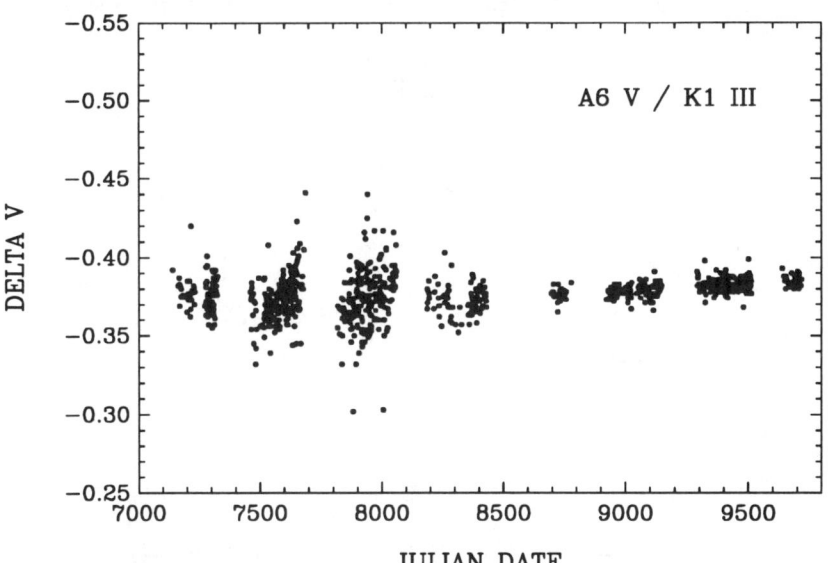

Figure 10. Differential magnitudes of the pair 27/28 Leo Minoris from 16-inch observations. The evolution of the observed scatter provides an eight-year history of the external precision of this telescope.

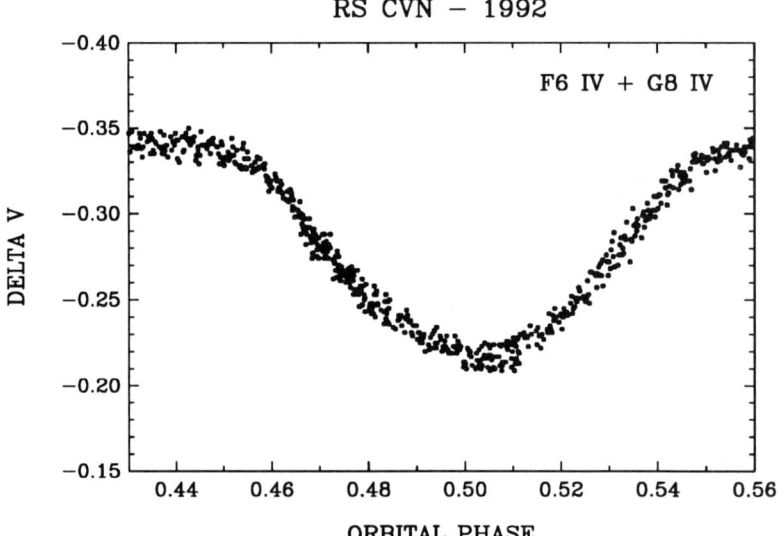

Figure 11. The 12-hour secondary eclipse of the spotted star RS CVn observed with the precision photometer on the 16-inch telescope. Asymmetries in the shape of the light curve helped map the distribution of starspots on the surface of the G8 IV star.

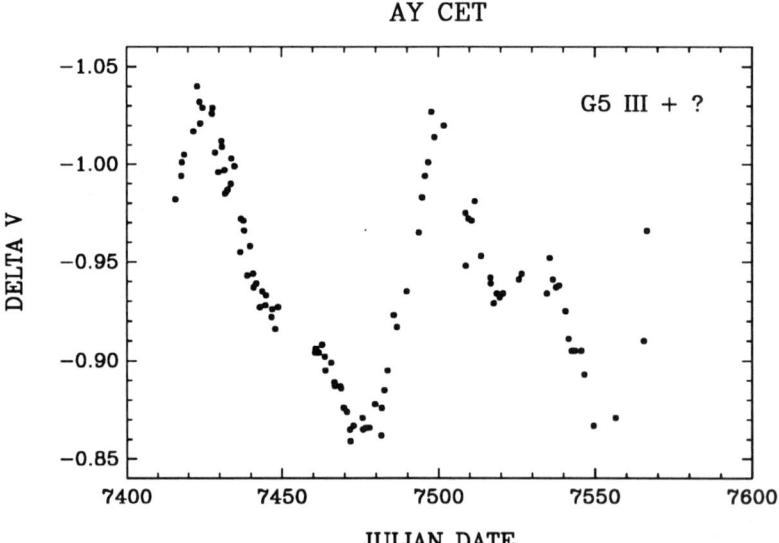

Figure 12. Two 77-day rotation cycles of the spotted G5 III star AY Ceti from the 16-inch APT. The formation of a new spot is evident on the second rotation.

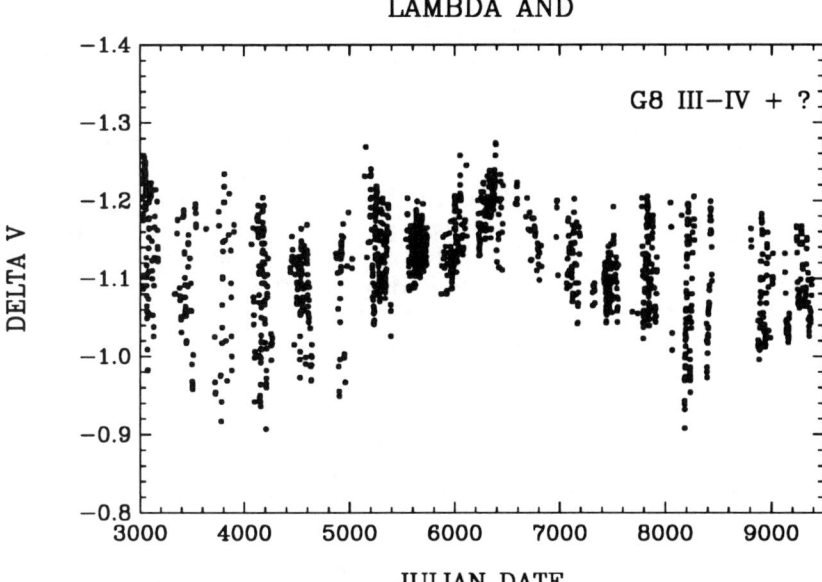

Figure 13. Seventeen years of manual and APT photometry of the spotted star λ Andromedae. The long-term variation in mean brightness suggests an 11-year starspot cycle.

separate nights of data for the 12-*hour* secondary eclipse of RS CVn. Every 4.8 days, the F6 IV primary star in RS CVn eclipses the spotted G8 IV secondary. Eaton et al. (1993) used the slight asymmetry in the light-curve to map the distribution of spots on the G8 star. Two 77-*day* rotation cycles of the spotted G5 giant AY Ceti are shown in Figure 12. As this star rotates, large starspots are carried into and out of view, resulting in the observed light variations. Tidal forces in the SB1 binary system sustain the rotation speed necessary to generate these large spots via a strong magnetic dynamo. On the second rotation, a secondary minimum appeared that revealed the emergence of a new starspot on the opposite hemisphere of the star from the spot causing the deep minimum. A 17-*year* light curve of the spotted star Lambda Andromedae is shown in Figure 13. The last 12 years of data were taken with two of the automatic telescopes on Mt. Hopkins. The vertical spread in each yearly data group is caused by light variations due to rotation of starspots. The long-term variation in mean brightness suggests a starspot cycle of 11 years (Henry et al. 1995), remarkably similar to the 11-year sunspot cycle.

4. The Smithsonian/Tennessee State University 30-inch APT

The desire to search for slight luminosity variations in solar-type stars as they go through their magnetic cycles required a further improvement in the precision of robotic telescopes. We know from space-based observations that the total solar irradiance varies by 0.001 mag in phase with the sunspot cycle (Fröhlich et al. 1991), although, historically, the Sun may have undergone somewhat larger brightness excursions (Baliunas & Jastrow 1990). The measurement of such small changes in other stars from ground-based APT's requires larger telescopes, longer integrations, extremely stable photometers, carefully planned observing strategies, and scrupulous quality-control measures.

The SAO/TSU 30-inch APT (Figure 14) represents a collaborative effort between the Center for Astrophysics, the Mt. Wilson Institute, and TSU to measure these small luminosity changes in solar-type stars and to correlate them with contemporaneous H & K measurements from Mt. Wilson (Baliunas et al. 1995). The 30-inch APT was constructed by Rettig Machine Shop of Redlands, California from a design by Boyd. Star Instruments produced the optics, and Boyd built the control system and precision photometer, a duplicate of the one operating successfully on the 16-inch APT. A year of development and testing finally culminated in April 1993 with the initiation of high-precision observations. Since then, over 11,000 group observations of solar-type stars have been accumulated.

The observations are made in Strömgren b and y passbands in a quartet group sequence: A,B,C,D,A,B,C,D,A,B,C,D. The use of three comparison stars (A,B,C) per program star (D) increases our chances of finding a comparison star constant to 0.001 mag or better. Sky observations were initially made at a single, central location for each group (as for the smaller APT's) after each D observation, but small systematic errors caused by gradients in sky brightness across the group when the moon was up caused us to revise the group sequence to make sky observations next to each star. Standard star observations are made and reduced each night, and nightly extinction and long-term mean transformation coefficients are applied in the differential reductions of the program star data, just as for the 16-inch telescope. On the best photometric nights, the internal precision ranges from about 0.0010 to 0.0015 mag, compared to 0.0013 mag predicted for the sum of photon and scintillation noise. Therefore, like the 16-inch APT since its precision photometer upgrade, the 30-inch operates at the theoretical limit of internal precision for its aperture and the integration times used. The data are cloud filtered at 0.005 mag (3-sigma). However, since the highest possible precision is desired for observing solar-type stars, only data on demonstrably photometric nights (i.e. nights on which the standard star observations give a good solution for all coefficients) are used in further analysis. These selected data, therefore, have not only survived the 0.005 mag cloud filter but have all been reduced with nightly extinction coefficients determined on good photometric nights.

In addition to standard star observations, other quality control observations are scheduled automatically each night to verify that all components of the telescope/photometer system are working properly. One of the most useful of these checks is the Fabry scan. Originally designed to be executed occasionally to verify that the Fabry lens was properly positioned, this test also proved useful for detecting problems with focus, CCD centering, optical alignment, filter wheel vignetting, and telescope drive systems. To run a Fabry scan, the telescope centers a star and then places it just outside the diaphragm. The telescope then steps the star through the center of the diaphragm in right ascension while the photometer takes an integration at each step. The star is recentered, and a similar declination scan is performed. The results of three such scans taken

Figure 14. The Smithsonian/Tennessee State 30-inch APT. Techniques for millimagnitude photometry of solar-type stars were developed on this telescope.

on different nights are illustrated in Figure 15. The top panel shows that the telescope is focused, that it is centering stars properly in the diaphragm, and that the Fabry lens is positioned properly to give a flat signal across the diaphragm. The middle panel indicates that a focus adjustment is needed since the wings of the scan are broadened. The bottom panel reveals that the star was not precisely centered in the diaphragm because the two scans are slightly offset from each other horizontally. Besides the Fabry scan, other tests are run each night to determine the deadtime of the photometer and to calibrate the neutral density filters (and thus to confirm that the filter wheels are functioning properly). All the standard star and quality control observations take less than 10% of the available observing time each night, and so they provide a detailed check on the quality of the night and the operation of the equipment at very low overhead.

The external precision of the 30-inch APT, as compared to the 10-inch and 16-inch telescopes, is shown in Figure 16. In all three panels, differential V magnitudes of the constant star pair HD 125451 / HD 124570 are plotted at the same scale for a recent observing season in which all three telescopes made observations of the constant pair. The external precision of the 30-inch APT is 0.0011 mag, an order of magnitude improvement over the early APT's. Furthermore, the *external* errors in typical group observations for the 30-inch are the same as the mean *internal* errors. Thus, the temperature stabilization of the photometer and the use of nightly extinction coefficients have succeeded in minimizing the external errors.

To demonstrate the capabilities of the 30-inch APT operating at this level of precision, we plot three rotation cycles of the G7 V star HD 152391 in Figure 17, phased together on its 11.5-day rotation period (determined from this photometry). This star exhibits a clear magnetic cycle of 10.9 years as deduced from its Ca II H & K emission (Baliunas et al. 1995). Even though the amplitude of its light variability over a rotation is only 3%, the high precision of the 30-inch APT data still produces a clean light curve.

Seasonal mean magnitudes of our solar-type stars will be searched for evidence of long-term luminosity changes associated with magnetic cycles. Because roughly 50 APT group observations are obtained per star per season, each with an external precision of approximately 0.0015 mag, the formal uncertainties of the seasonal mean magnitudes will be around 0.0002 mag. Table 1 presents 1993 and 1994 seasonal means in the combined Strömgren $b + y$ passbands for the old G1 V star HD 126053. The three comparison stars (A,B,C) appear to be extremely stable since the C-A, C-B, and B-A differential magnitudes of these comparisons all agree from one season to the next to within 0.0001 mag. The program star differential magnitudes with respect to the three comparison stars (D-A, D-B, and D-C) all show an increase in the brightness of HD 126053 (star D) of 0.0002 or 0.0003 mag from 1993 to 1994. This small amplitude is just the amount the Sun's brightness would be expected to change over one year as it progresses through its magnetic cycle. With sufficiently stable comparison stars (admittedly not easy to find), it appears that we can measure these small luminosity changes in solar-type stars.

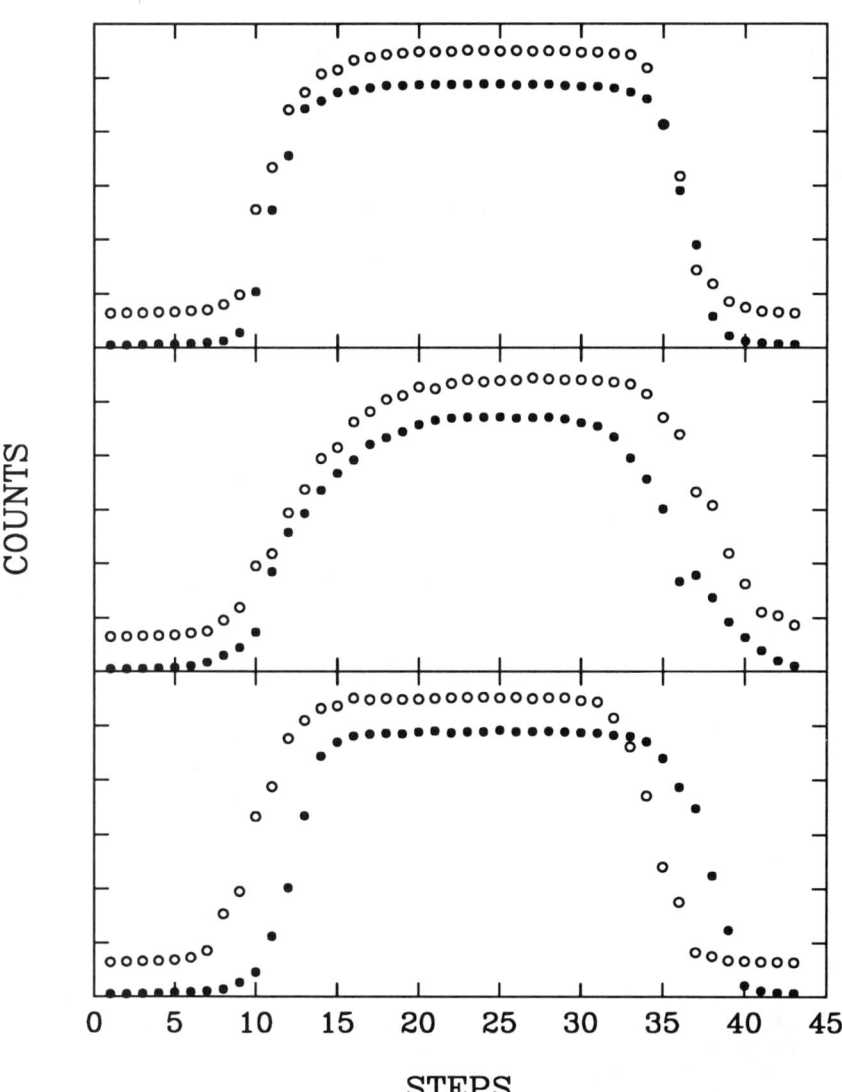

Figure 15. Fabry scans from the 30-inch APT. Open circles represent the scans in right ascension (shifted upward for separation) while closed circles represent the declination scans. The top panel (JD 2449746) implies a focused telescope with good centering. The middle panel (JD 2449745) shows the telescope is out of focus. The bottom panel (JD 2449751) indicates a slight centering adjustment is needed.

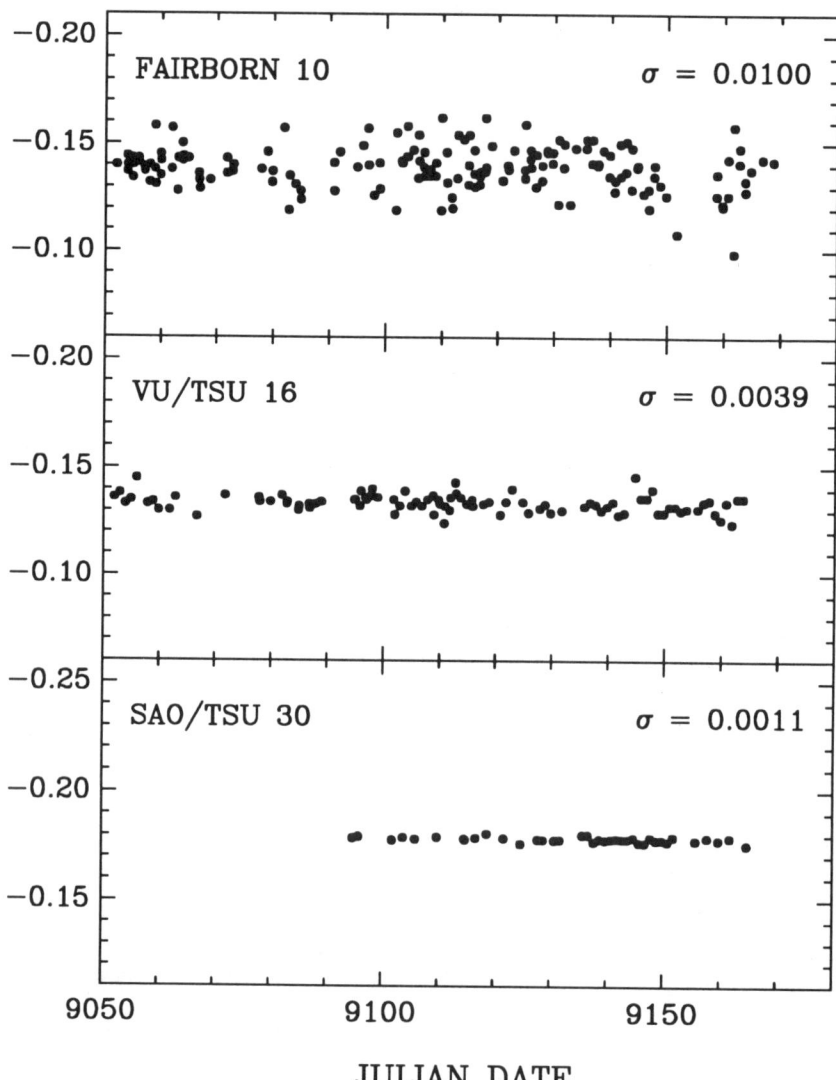

Figure 16. The external precision of the 30-inch APT compared to the 10-inch and 16-inch telescopes. The developments described in this paper have resulted in an order of magnitude improvement in the precision of robotic telescope photometry.

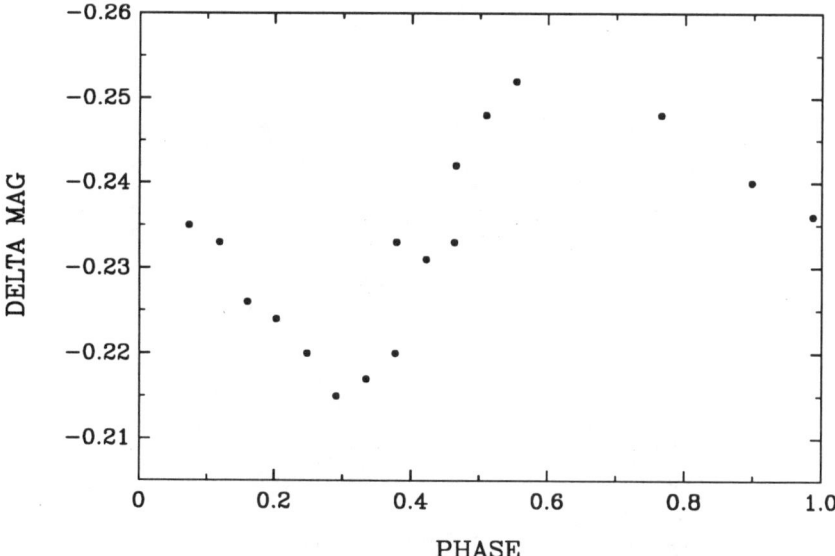

Figure 17. Light curve of the G7 V star HD 152391 obtained with the 30-inch APT. Three rotation cycles are phased together on the 11.5 day rotation period. The high precision of the 30-inch photometry results in a clean light curve even though the amplitude of variability is only 3%.

5. The Tennessee State University/Smithsonian 32-inch APT

A fourth telescope operated by TSU will go into operation at Fairborn in 1995. This 32-inch APT (Figure 18) will be dedicated to the measurement of luminosity changes in main-sequence stars that are very close to the Sun in both mass and age, a sub-group of the solar-type stars we call the solar duplicates. Construction of this telescope was originally contracted to AutoScope Corporation for location on Mt. Wilson. However, the failure of AutoScope to produce a working telescope (Henry 1995) forced us to relocate the telescope at Fairborn and have Boyd bring it into operation.

Because of the extremely small amplitude of the expected light variations of solar duplicates, Boyd has designed and constructed a two-channel precision photometer especially tailored for our requirements. A dichroic filter separates the Strömgren b and y passbands so that separate EMI 9124 QB photomultiplier tubes can measure the two colors simultaneously. Because scintillation noise is the principal source of error in the 30-inch telescope, we expect to improve the precision further with this telescope by doubling all integration times. The two-channel photometer allows us to do this with no loss of observing efficiency. We anticipate that the precision of a nightly group observation will be less than 0.001 mag.

Figure 18. The Tennessee State/Smithsonian 32-inch APT. A new two-channel photometer designed for this telescope will make submillimagnitude observations of solar duplicates.

Table 1. Seasonal mean magnitudes for the old G1 V star HD 126053

Pair	1993 Seasonal Mean	Sigma
D-A	0.0882	0.0015
D-B	-0.7320	0.0018
D-C	-0.1095	0.0014
C-A	0.1977	0.0014
C-B	-0.6226	0.0018
B-A	0.8202	0.0019

Pair	1994 Seasonal Mean	Sigma
D-A	0.0880	0.0016
D-B	-0.7323	0.0014
D-C	-0.1097	0.0017
C-A	0.1976	0.0016
C-B	-0.6226	0.0017
B-A	0.8203	0.0013

Acknowledgments. It is a pleasure to acknowledge scientific collaborators Douglas S. Hall, Sallie L. Baliunas, Joel A. Eaton, and Frank C. Fekel for their part in the development of precision robotic photometry. This effort has been driven by stimulating scientific programs with unquenchable needs for more, better, and cheaper photometry. Thanks are also due to Irwin Shapiro, director of the Center for Astrophysics, for his continued support of the Fairborn Observatory site on Mt. Hopkins and to Mike Busby, director of the Center of Excellence in Information Systems, for providing a supportive environment for this work over the past several years. Of course, none of this would have been possible without the tireless expertise of Lou Boyd at Fairborn. These programs have also been supported for several years at Tennessee State University by the National Aeronautics and Space Administration and by the National Science Foundation, most recently through NASA grant NAG 8-1014 and NSF grant HRD-9104484.

References

Baliunas, S. L. et al. 1995, ApJ, 438, 269
Baliunas, S. L. & Jastrow, R. 1990, Nature, 348, 520
Boyd, L. J., Genet, R. M., & Hall, D. S. 1984, IAPPP Comm. No. 15, 20
Cristina, V.-C., Donahue, R. A., Soon, W. H., Baliunas, S. L., & Henry, G. W. 1995, PASP, in press

Eaton, J. A., Henry, G. W., Bell, C., & Okorogu, A. 1993, AJ, 106, 1181

Fröhlich, C., Foukal, P. V., Hickey, J. R., Hudson, H. S., & Willson, R. C. 1991, in The Sun in Time, eds. C. P. Sonett, M. S. Giampapa, and M. S. Matthews (Tucson: University of Arizona Press), p. 11

Genet, R. M., & Hayes, D. S. 1989, Robotic Observatories, (Mesa: AutoScope)

Hall, D. S. & Henry, G. W. 1993, in Stellar Photometry - Current Techniques and Future Developments, eds. C. J. Butler & I. Elliott (Great Britain: Cambridge University Press), p. 205

Henry, G. W. 1995, IAPPP Comm. No. 57, 74

Henry, G. W., Eaton, J. A., Hamer, J., & Hall, D. S. 1995, ApJS, in press

Young, A. T. 1974, in Methods of Experimental Physics, Vol. 12A (Astrophysics, Optical and Infrared), ed. N. Carleton (New York: Academic Press), p. 101

Young, A. T. 1991, private communication

Young, A. T., Genet, R. M., Boyd, L. J., Borucki, W. J., Lockwood, G. W., Henry, G. W., Hall, D. S., Smith, D. P., Baliunas, S. L., Donahue, R., & Epand, D. H. 1991, PASP, 103, 221

Variable Stars in the Hertzsprung-Russell Diagram

Douglas S. Hall

Dyer Observatory, Vanderbilt University, Nashville, Tennessee 37235

Abstract. A classification scheme based on the physical mechanisms that cause a star to be variable is proposed. The six main categories are (I) eclipsing or extrinsic, (II) pulsation, (III) rotation, (IV) long-term magnetic cycles, (V) supergranulation, and (VI) transient phenomena. Seven stars are given as examples, "classified" according to the mechanisms that produce their multiperiodic variability. It is shown how the 130+ GCVS variable-star types could be re-assigned to this new classification scheme, but quite a few GCVS types cannot logically fit, for obvious reasons which are explained.

The benefit of plotting the domains of the various known types of variable stars in the H-R diagram is discussed, and the problems. The ultimate purpose should be to map out the domains where various physical mechanisms for variability are operative. Certain variability mechanisms cannot meaningfully be indicated. The full extent of some domains may never be established because of rapid evolution through, for example, the Hertzsprung gap. In some cases, all stars within a certain domain *will be* variable. In other cases, variability within that domain is *possible* but depends on one or more parameters besides surface temperature and absolute luminosity, such as strong magnetic field or rapid rotation.

Important questions to answer are identified. (1) Does a certain mechanism, tentatively hypothesized, actually occur or not? Examples: supergranulation and magnetic cycles. (2) Can a recently proposed mechanism be confirmed with greater confidence? Example: large starspots. (3) Can domain boundaries be defined more sharply? Example: the pulsation/radiation-pressure interface. (4) Where domains overlap, can we devise methods to determine which one is operating in a given star? (5) What mechanisms cause the variability in some of the still puzzling GCVS types? Example: the Orion variables. (6) Are there totally new mechanisms for variability, which we don't even know of? Example: the puzzling F0 V variables.

Three specific observing programs designed to answer some of these "important questions" are proposed. (1) Demonstrate that there are no G-type or K-type giants which pulsate; 17 variables are identified for observation. (2) Verify that a Rossby number less than 2/3 *guarantees* large starspots; 28 apparently non-variable stars are identified for observation. (3) Discover whether the puzzling F0 V variables represent a totally new mechanism; a list of 10 such variables is provided.

1. Introduction

My motivation for this paper is to stimulate excitement for continued study of variable stars. It seems to me that variable star research could profit from some rejuvenation because of some misperceptions held by many astronomers. Some feel that the field is nearing saturation if only because so many different variable stars and different types of variable stars have been discovered and catalogued already. Some feel that most of the astrophysically important questions which are needed to understand stellar variability or which can be learned from a study of variable stars, have been answered/learned already. These perceptions stem from the fact that our current body of knowledge, what we do know about variable stars and stellar variability in general, is poorly organized and confusing. This confusion has made it difficult to see clearly all that we do know and, the other side of the coin, what we don't know but need to.

The acknowledged final word on variable stars is the "General Catalogue of Variable Stars" (GCVS). The five-volume Fourth Edition, the first four volumes of which have been published (in 1985, 1987, and 1990), lists approximately 28,450 different variable stars discovered and formally named by 1982, and over 130 different types. As new variable stars are discovered and formally named, they are announced in a numbered "Name List of Variable Stars", each one an issue of the I.A.U. Commission 27 "Information Bulletin on Variable Stars". The latest one, as I write this paper, was the 71st Name List, which appeared as I.B.V.S. No. 3840 (Kazarovets et al. 1993). The exact number of types varies somewhat from one volume of the GCVS to the next and new types occasionally are introduced in one of the Name Lists. Moreover, a given variable star can be assigned a hybrid or combination type. Separate from the variable stars formally named in the GCVS or in the Name Lists, are the suspected variables, the most recent compilation of which is the "New Catalogue of Suspected Variable Stars". The New Catalogue is the third such catalogue of suspected variables but it changed completely the numbering system used in the first two editions.

2. A Proposed Better Way to Organize Variable Stars

It would be more instructive, it appears to me, if variable stars were classified according to the physical mechanism which causes their light to be variable in time. The GCVS nods in this direction with its establishment of five "classes" of variable stars: eruptive, pulsating, cataclysmic, close binary eclipsing systems, and x-ray sources; but the overall classification is confusing and inadequate, because only some of these five main classes meaningfully represent distinct physical mechanisms for variability in light output. A few examples. (1) Emission of x-rays *per se* says nothing about variability in light output. If wavelength region outside the optical is to be considered important, why isn't there also a class to recognize the radio stars or the infrared stars? (2) The close binaries are sub-classified in several different, parallel schemes, most of them completely unrelated to variability *per se*. Only the old EA,EB,EW scheme directly addresses variability but, except for the E which means "eclipsing", that scheme does not cleanly identify different mechanisms for producing light variability. (3) It is clear that "eruption" is not a physically well defined mechanism, let alone one

Table 1. Physical mechanisms for stellar variability.

I. Eclipses
    A. in a binary star system
    B. by dust external to the star
    C. occultation by a nearer star, planet, asteroid, moon
    D. gravitational microlensing by a nearer star
II. Pulsation
    A. radial or non-radial
    B. singly periodic or multiply periodic
    C. p-modes or g-modes
III. Rotation
    A. non-spherical shape = the ellipticity effect
    B. longitudinally asymmetrical surface brightness distribution
        1. by the reflection effect in a binary system
        2. spots at the magnetic poles in strongly magnetic stars
        3. cool spots in strong-dynamo convective stars
        4. hot spots
IV. Magnetic cycles akin to the 11-year sunspot cycle
V. Supergranulation
VI. Eruptive = transient release of energy or movement of matter
    A. involving the entire star = supernova
    B. from the stellar surface
        1. a nuclear explosion
        2. a magnetic explosion = flare
        3. outflow of matter caused by rapid rotation
        4. outflow of matter by another mechanism, e.g., radiation pressure
    C. infall of matter = accretion
        1. from the other star in a binary system
        2. from a circumstellar disk around a single star

which meaningfully ties all of the different types of "eruptive variables" together and distinguishes them from the "cataclysmic variables". And (4) a few types within a class don't logically belong there at all. For example, there never has been anything "eruptive" about the photometric behavior of the RS CVn-type binary stars.

Table 1 outlines a classification scheme based on the physical mechanism which causes a star's light to be variable in time. It embraces variability of any amplitude, in any wavelength region, on any time scale shorter than a core-driven evolutionary time scale, and periodic or irregular or transient.

Group I. Variability by the mechanisms in this group is not intrinsic to the star itself. (A) In eclipsing variables, one star does it to the other. (B) It may be that clumps of dust orbiting around an extremely young star which still has its circumstellar disk can cause that star to vary in light. This has been proposed as the mechanism for one component of the irregular variability observed in T

Tauri-type and similar stars, but that is not firmly established. (C) A star's light will vary, i.e., dim, when it is occulted by a solar system object. Stars which vary in light in this way have never been named as variable stars nor included in variable star catalogues, and shouldn't be, but they are "variable stars" in the most general sense of the term. (D) Quite recently a few stars have been observed to vary in light, this time an increase, as a result of gravitational microlensing. The "lens" is thought to have been a star, possibly one of very small mass, lying between the Earth and the "variable star" in question. As with the occulted stars, these microlens variables are not being named formally as variable stars.

Group II. Pulsation is perhaps the most distinct, physically specific mechanism for intrinsic stellar variability. Many of the "types" of pulsating variables defined by the GCVS seem to have been defined in accordance with the star's evolutionary state or population type or spectral type or luminosity class. It would be more instructive, *vis à vis* the underlying physics, to ask if the pulsation is radial or non-radial, periodic or irregular, singly periodic or multiply periodic, powered by p-mode or g-mode pulsations, triggered by the first or second helium ionization zone or the hydrogen ionization zone, etc.

Group III. (A) Rotation will produce variability in luminous output if the star's shape is non-circular in the plane perpendicular to its rotation axis. This will occur naturally only in a binary system, where one star is rendered prolate in shape by tidal interaction with its companion, not in a single star, where rotation makes the star oblate. (B) Rotation can produce variability in a spherical star if, for any reason, the surface brightness has a longitudinally asymmetric distribution. Four different physical mechanisms for possible asymmetry are listed. Some, like the reflection effect and the hot spots resulting at gas stream impact sites, will arise only in binaries. The two types of magnetic spots will be intrinsic to a particular star, but that star may or may not be in a binary system.

Group IV. This mechanism refers to the recently established long-term variability in the Sun's mean bolometric luminosity and in the mean luminosity of several other stars and types of stars (Hall 1990, 1991a, Applegate 1992, Henry et al. 1995). In the Sun's case, the variability appears to be in phase with the 11-year sunspot cycle and hence promises to be understood by the same underlying physical mechanism, whatever that is. In the case of the other stars, argument by analogy suggests the same mechanism. To my knowledge, all stars in which this sort of long-term variability has been established are previously designated variable stars, with their variability attributed to a different mechanism. Consequently, you will not find this variability mechanism included as one of the 130 or more mechanisms listed in the GCVS.

Group V. The supergranulation referred to here is taken to be the sort of phenomenon discussed by Schwarzschild (1975). It frankly has not been established that this mechanism can or actually does produce variability in any part of the Hertzsprung-Russell diagram, but, on the other hand, it has never been demonstrated that it does not. Because it works to account for variability in one of the least understood regions of the H-R diagram, where the brightest and coolest supergiants and the puzzling variables of GCVS type LC and SRC lie, I am persuaded to allow for it as a possible physical mechanism distinct from all of the others.

Group VI. Finally we come to the general category of <u>transient phenomena</u> which produce light variability. The supernovas stand alone in their own subgroup (A), unique because they involve a one-time total anihilation of a star. The last two subgroups distinguish between (B) explosions on or eruptions from the surface of a star and (C) infall *onto the surface* of a star, both of which can be repetetive.

In my scheme, a given variable star would be "classified" according to the mechanism or mechanisms that produce the variability or components of variability which are observed. A few examples would be

V711 Tau = HR 1099 would be (III-A)+(III-B-3)+(IV)+(VI-B-2); III-A because of its small ellipticity effect, III-B-3 because of its starspot variability, IV because of the long-term cycle in its mean brightness, and VI-B-2 because of its white-light flares.

Algol would be (I-A)+(III-A)+(III-B-1)+(III-B-3); I-A because it is eclipsing, III-A because of its small ellipticity effect, III-B-1 because of its reflection effect, and III-B-3 because of the secondary star's recently detected starspot variability.

DQ Her would be (I-A)+(II)+(III-B-4)+(VI-B-1)+(VI-C-1); I-A because it is eclipsing, II because its white dwarf pulsates, III-B-4 because the hot spot on its accrection disc rotates, VI-B-1 because of the nova explosion, and VI-C-1 because of the flickering caused by unsteady mass transfer.

the Crab pulsar would be (III-B-2) because of its oblique, rotating, light-emitting magnetic poles.

R CrB would be (II)+(VI-B-4); II because it pulsates slightly, and VI-B-4 because of the ejection of dust, probably by radiation pressure.

T CrB would be (II)+(III-A)+(VI-C-1); II because the late-type giant is a semi-regular pulsator, III-A because of the ellipticity effect, produced mostly by the late-type giant, and VI-C-1 because transient episodes of mass transfer and subsequent accretion energy cause the recurrent nova outbursts.

RR Lyr would be (II)+(III-B-2)+(IV); II because it pulsates, III-B-2 because of the Blazhko effect, explained via the oblique rotator model, and IV because of long-term (probably magnetic) cycles which have been observed.

It might be illuminating to indicate how the various GCVS variable star types could be re-assigned to my classification scheme. This is shown in Table 2. For a complete list of GCVS types I have taken the one on page 362 of the third volume of the Fourth Edition but added types ZZB, ZZO, and SNII which were mentioned in other volumes. Note that I have filled up my mechanism I-B with several GCVS types even though, as explained earlier, it has not been firmly established that clumps of dust orbiting in a circumstellar disk around these young stars really does account for their variability. Similarly, I have placed the LC and SRC variables in my Group V, even though supergranulation

Table 2. GCVS types assigned to the physical mechanism scheme.

I. Eclipses
   A. E, EA, EB, EW
   B. IN, INA, INAT, INB, INS, INSA, INSB, INST, INT
   C. no GCVS type
   D. no GCVS type

II. Pulsation
   ACYG, BCEP, BCEPS, CEP, CW, CWA, CWB, DCEP, DCEPS,
   DSCT, DSCTC, LB, LC, M, PVTEL, RR, RRAB, RRC, RV, RVA,
   RVB, SR, SRA, SRB, SRC, SRD, SXPHE, ZZ, ZZA, ZZB, ZZO,
   ACVO, RCB, (B)

III. Rotation
   A. ELL, XP
   B. 1. XPR
      2. ACV, SXARI, PSR, ACVO
      3. RS, BY, FK, IN, INA, INAT, INB, INT, IT, INS, INSA,INSB,INST
      4. AM, XPM

IV. Magnetic cycles
   no GCVS type

V. Supergranulation
   LC, SRC

VI. Eruptive
   A. SN, SN-I, SN-II
   B. 1. N, NA, NAB, NB, NC, XB
      2. UV, UVN
      3. GCAS
      4. WR, SDOR, RCB, NC, ZAND, XJ, SRD
   C. 1. UG, UGSS, UGSU, UGZ, ZAND
        X, XB, XBPR, XBI, XF, XFND, XI, XJ, XM
        XN, XND, XNDB, XNGP, XP, XPM, XPR, XR, XRM
      2. FU, (YY)

Table 3. GCVS types not assigned to the physical mechanism scheme.

I, IA, IB – poorly studied, irregular variables, very inhomogeneous group
IS, ISA, ISB – irregular variables, many probably mistakenly attributed
LSA, LSB – irregular variables, many probably mistakenly attributed
NL – inhomogeneous group, poorly studied, probably will end up being reclassified into other types
NR – inhomogeneous group, variable by a variety of different physical mechanisms
GS, PN, RS, WD, WR – a classification based on stellar physical characteristics, not related to photometric variability
AR, D, DM, DS, DW, K, KE, KW, SD – a classification based on degree of Roche lobe filling, not related to photometric variability
BLLAC, GAL, QSO – galaxies, not stars
CST – non-variable stars
* – unique
L:, S: – unstudied variable stars

---

has not been established as the principal mechanism for their variability. That still leaves some of my mechanisms (I-C, I-D, and IV) empty, with no GCVS counterpart.

Quite a few GCVS types do not logically fit into my classification scheme, for obvious reasons which are explained in Table 3. Several are acknowledged by the GCVS to be comprised of inhomogeneous samples of poorly studied stars mistakenly attributed and likely to be reassigned to another type upon further investigation. Others were not even intended by the GCVS to represent mechanisms for variability at all.

3. Variable Stars in the Hertzsprung-Russell Diagram

Since my student days, I have been intrigued with the picture of an H-R diagram showing the domains of the various known types of variable stars. This seemed to "pull it all together". By now there have been several versions of such H-R diagrams published in books (Glasby 1969, figure 13; Strohmeier 1972, figure 24; Cox 1980, figure 3.1; Petit 1982, figure 3; Duerbeck & Seitter 1982, figures 1 and 2; Hoffmeister et al. 1985, figure 163) not counting numerous astronomy text books. All of these are, frankly, deficient in one or more ways. None of them includes all of the known variable star types, either because they pre-date the discovery of many of the types or they chose to include only some of the types, for example, only the pulsating variables. Several of them are simply reproductions of an earlier one. Some contain obvious errors. It is frustrating that none of them has provided bibliographic references to the specific source material used to plot the domains. That has made it difficult for anyone else, like me, to update or amend or add to the existing diagram without having to reconstruct it entirely *de novo*.

The ultimate purpose of an H-R diagram with variable star domains plotted should be to map out the domains where various physical mechanisms for variability are operative. The flip side of this is to enable us to identify regions of the H-R diagram where stars are reasonably guaranteed not to be variable. Variable star observers looking for reliably constant comparison stars would appreciate this, even if no one else was interested.

Certain variability mechanisms cannot meaningfully be indicated. Stars variable by any of the extrinsic mechanisms in my Group I should be excluded. Variability in binaries by the ellipticity effect (III-A) or the reflection effect (III-B-1) and possibly hot spots produced by mass transfer (III-B-4 and VI-C-1) can't meaningfully be shown. And it would be difficult to define where the supernova precursors lie. Moreover, whenever the variable star is in a binary, care should be taken to plot the spectral type and absolute magnitude of the component which is the intrinsic variable, not a composite of the two components. Once these mechanism domains are delineated, their significance needs to be qualified carefully.

The full extent of some domains may never be established simply because evolutionary tracks carry stars so quickly through some regions of the H-R diagram that we probably never will see them, even though they may be variable as they go. The so-called Hertzsprung gap is the best known example.

In some cases, all stars within a certain domain *will be* variable. The clearest example would be the famous pulsational instability strip, which includes the Cepheids, the RR Lyrae variables, and the δ Scuti variables. Another example would be the extremely luminous supergiants (GCVS types WR, SDOR, yellow supergiant SRD like ρ Cas, and especially luminous ACYG) which might all vary by the mechanism of irregular outflow of matter due to radiation pressure (IV-B-4) if they are sufficiently luminous, though the "all" has not yet been firmly established.

In other cases, however, variability within that domain is *possible* but depends on one or more parameters besides surface temperature and absolute luminosity. A clear example of this is the narrow domain where the so-called magnetic variables (a.k.a. peculiar A-type stars, a.k.a. chemically peculiar stars) reside (III-B-2). These consist of the ACV variables, if they are of spectral type A, and the SXARI variables, if they are of spectral type B. In either case, however, the star must have an extraordinarily strong permanent magnetic field; otherwise they will be non-variable, normal, i.e., non-peculiar A-type stars. Another clear example is the stars which are variable due to rotational modulation of large starspot regions on their surface (III-B-3). It has been shown (Hall 1991b, 1994) that the large starspots required for measurable variability result from a certain critical combination of B-V, absolute magnitude, and rotation period. Note that the last of these is the extra parameter, over and above the two needed to plot a star in the H-R diagram, required to assure variability. A third example would be the GCAS variables, where variability results from outflow of matter caused by rotation beyond the centrifugal limit (VI-B-3). The domain is very narrow in both spectral type and luminosity class. Although stars at this upper end of the main sequence do tend to be very rapid rotators, it is not clear if all of them rotate rapidly enough for centrifugal break-up. At the very least one will find those in close binaries (perhaps as many as half of all B-type main-sequence stars) have been spun-down and do not rotate rapidly enough to trigger the GCAS variability mechanism.

Of the published H-R diagrams showing variable star domains, the ones by Duerbeck & Seitter (1982) and by Hoffmeister et al. (1985) are the most complete and informative with respect to the questions posed in the paragraphs above. Pointing to the second of these, let me suggest the few additions, based largely on what we already know, which would make it nearly perfect. (1) The variable Wolf-Rayet stars (GCVS type WR), which may vary by my mechanism VI-B-4, are not shown. They could extend, to the left of the SDOR variables, the domain covered by that mechanism, namely, mass loss by radiation pressure. (2) The really luminous stars of GCVS type SRD, for which ρ Cas could be considered a *de facto* prototype, are not shown. They could extend, to the right of the SDOR variables, the domain controlled by mechanism VI-B-4. (3) The ACYG and PV Tel variables are not shown. They could be useful in showing that the domain of early-type pulsators might extend continuously further up in luminosity from the BCEP variables. (4) The ACVO variables, magnetic variables which also pulsate slightly, are not shown. They could be useful in showing that the domains of the magnetic variables and the δ Scuti variables do formally overlap in one place. (5) Only BY Dra variables are plotted to indicate mechanism III- B-3 but my Table 2 shows how many other GCVS types (single and binary, pre-main-sequence, main-sequence, and post-main-sequence) vary by that same mechanism. In actuality, starspot variability populates a vastly larger region of the H-R diagram. The bottom edge is the zero-age main sequence; the left edge is the vertical line representing F5; the right edge is the Hayashi track or, almost equivalently, the post-main-sequence evolutionary track of a star around one solar mass; and the top edge is the locus of luminosity class III (giant) stars.

4. The Next Important Questions to Answer

So that variable star research can be as instructive as possible, *vis à vis* teaching us new astrophysics, we should focus on the underlying physical mechanisms which cause variability. Incisive questions will be along the following lines. (1) Does a certain mechanism, which has been tentatively hypothesized, actually occur in nature or not? (2) Can a relatively recently proposed mechanism for variability be confirmed with greater confidence? (3) Can the boundaries of the domains (in the H-R diagram or some other parameter space) for the various variability mechanisms be defined more sharply and, where domains overlap, can methods be devised to determine which one is operating in a given star? (4) What is the mechanism that causes the variability in some of the still puzzling GCVS types? (5) Are there any totally new mechanisms for variability, which we don't even know of?

Does the supergranulation of Schwarzschild (1975) explain some or all of the variability in certain late-type supergiants, with the time scale of the variability corresponding to the lifetime of a characteristic supergranule, as he suggested? The most likely candidates for finding this mechanism at work would be the GCVS LC type ("irregular variable supergiants of late spectral type") and SRC type ("semi-regular variable late type"). Less likely candidates would be the

LB, SRA, and SRB types, late-type irregular or semi-regular variables which are giants instead of supergiants.

Does the SRD type ("semi-regular variable giants and supergiants of spectral types F,G,K") throw together variables in which a number of very different mechanisms cause the variability? I suspect that, if they are broken down in the following way, their variability can be understood reasonably well. (1) Those of spectral types F, G, K and luminosity classes Ia, Ia-0, 0 will prove to vary by my mechanism VI-B-4, with the "hypergiant" ρ Cas being useful as a proto-type, as discussed in the last paragraph of the preceeding section. (2) Those of spectral types F, G, K and luminosity classes II and Ib will prove to be pulsating (my mechanism II) and hence be similar to, or perhaps identical with, pulsating variables of the CEP, CW, or RV type. (3) Those of spectral type F and luminosity class III will also prove to be pulsating but in this case be similar to variables of the DSCT or RR type. (4) Those of spectral type G or K and luminosity class III will prove not to be pulsating, with their variability resulting from an assortment of miscellaneous mechanisms other than pulsation, such as eclipses or starspots or the ellipticity effect.

Two of the domains have particularly fuzzy boundaries and may, in addition, overlap each other: pulsation (II) and radiation pressure (VI-B-4). First of all let me explain that, in the special case where radiation pressure expels *dust* rather than hot gas, we already have a pocket of VI-B-4 variables sitting inside the extensive domain of pulsating variables. The R CrB variables themselves, GCVS type RCB, illustrate this because many of them vary by both mechanisms. But let us focus on the overlap problem where radiation pressure expels hot gas. At early spectral type, relatively luminous variables of the PVTEL and ACYG type, generally thought to be pulsators, actually lie in the domain where most of the SDOR variables lie. At intermediate spectral type, we don't know how far down in the H-R diagram (how much lower in luminosity) the cousins of ρ Cas are found, so we don't know if their domain does or doesn't overlap with the most luminous of the pulsators, which would be the classical cepheids of longest period. At late spectral type, we don't even know what makes the supergiants vary: pulsation or radiation pressure (or supergranulation). So it is an understatement to say that the domain overlap problem is very serious here at the upper righthand corner of the H-R diagram. If variability by supergranulation proves to be real, then the domain overlap problem becomes far worse still.

What is the domain in which mechanism IV, magnetic cycles, operates to produce measurable variability on a decade-long time scale? This is perhaps the newest mechanism on any such list and, not surprisingly, its domain has not been defined. Because both these magnetic cycles and the large starspots responsible for mechanism III-B-3 are intimately related in the picture of a stellar dynamo, we can conjecture that their domains will be similar if not virtually indistinguishable. But the work of turning conjecture into fact does remain to be done.

What does cause the irregular variability seen in most of the extremely young, pre-main-sequence stars. The GCVS calls these stars "Orion variables" and uses the letter "N" in the several variable star types assigned to them, because they are generally associated with nebulosity. Earlier it was thought that clumpy dusty material orbiting in a circumstellar disk produces the irregular variability, by obscuration of the starlight, my mechanism I-B. Later, instances were found of infalling matter, from the circumstellar disk onto the star, producing hot spots which then produced variability by my mechanism III-B-4 or VI-C-2. Later still, it has been established that at least one component of the variability, semi-periodic in nature, results from rotational modulation of large dark starspot regions, my mechanism III-B-3. And it is quite reasonable to expect variability by mechanism IV, because all strong-dynamo stars are quite similar in interior structure, no matter what their evolutionary state. The question can be sharpened: do we need all of these mechanisms to explain the Orion variables, or can one of them, namely, the clumpy dust, be discarded?

The next three sections outline three specific observing programs designed to answer some of these "important questions".

5. Are There No G-type or K-type Giants Which Pulsate?

In section 4 it was proposed that variables of the GCVS type SRD might be understood as a collection of several very different types. Giants (luminosity class III) of spectral type G or K, it was proposed, are not pulsating but rather are varying by a variety of other (non-pulsational) mechanisms. To test this hypothesis, I compiled a list of all stars in the first three volumes of the Fourth Edition of the GCVS which have been classified spectral type K and luminosity class III. The total number was 55.

The clear majority of these (38/55 = 69%) had GCVS types which indicated non-pulsational variability mechanisms: (1) Ten were "cst", which means they are constant, not variable stars after all. (2) Ten were "RS" or "FKCOM", which means starspots, my mechanism III-B-3, are causing the variability. (3) Nine were eclipsing binaries. (4) Two were "ELL", i.e., ellipsoidal variables. (5) Five were "IN", what the GCVS calls the "Orion variables"; it was explained in section 4 that their variability mechanism (or mechanisms) still is not understood, but pulsation has never been suggested. (6) One was "RCB", of the R CrB type. (7) One was "ZAND", which most of us know better as the symbiotic variables.

The minority of these (17/55 = 31%) had GCVS types which indicated some sort of pulsation might be the variability mechanism. These are listed in Table 4. (1) One was "ISB", which means rapid irregular variable of intermediate or late spectral type. (2) Seven were "L, LB, or LC", where the "L" means slow irregular variable. (3) Nine were "SR, SRB, or SRD", where "SR" means semi-regular variable. Let me add that most of these (11 of the 17) were marked with a colon (:) to indicate the GCVS classification type was uncertain.

The finger is pointed at this remnant of 17. The hypothesis (there are no pulsating K-type giants) will have been confirmed if additional observation, photometry and/or spectroscopy, can demonstrate any of the following to be true. (1) They are of spectral type M, rather than K, in which case they might understandably be Mira-type variables or variables of type SRA or SRB, the low amplitude cousins of the Miras. (2) They are more luminous than luminosity class III, in which case they might be reasonable candidates for pulsating variables of the Cepheid, W Virginis, or RV Tauri type. (3) They are not variable. (4) The radial velocity changes or color changes normally associated with pul-

Table 4. K-type giants predicted to be not pulsating or not K giants.

GCVS name	GCVS type	GCVS spectral type	GCVS ampl.	GCVS mag. at max.
HK Aql	LB:	K5 III	1.0	8.9
AW CVn	SR:	K5 III	0.09	4.72
V538 Cas	ISB	K5 III	1.2	9.4
YZ Cen	SR:	K3 III	0.5	10.5
V418 Cen	LC	K4 II	0.8	8.7
V764 Cen	SRD	K2 III	0.29	8.84
T Cyg	LB:	K3 III	0.05	4.91
VW Dra	SRD:	K1.5 IIIb	1.0	6.0
V463 Her	SRD	K0 III	0.05	8.46
SY Leo	SRB	K8 III – M	0.9	12.2
ST Lyn	LB:	K0 IIIea	1.0	11.7
EF Mus	LB:	K1 III-II	0.8	8.0
κ Oph	LB:	K2 III	0.9	4.1
CK Ori	SR:	K2 IIIe:	1.2	5.9
RX Ret	SRD	G8-K0 III-II ep	2.1	9.1
TZ Sge	SRD:	K2 III	3.3	12.7
α Tau	LB:	K5 III	0.20	0.75

sation are found to be absent. (5) The presence of a different, non-pulsational, variability mechanism is demonstrated, examples being starspots, eclipses, or the ellipticity effect.

As I was preparing the talk which led to this paper, the excellent paper by Percy (1993) came to my attention. This work effectively confirms the hypothesis by approaching the question from a different angle. He photoelectrically monitored 49 of the more than 200 K giants in the "Yale Catalogue of Bright Stars" which are named or suspected variables, finding only two clearly variable and a few more marginally variable. Most of them, including the two which were clearly variable, he felt were most likely starspot variables. His conclusion was that, though he could not rule out the possibility that a *few* K giants are SR variables (and by that he implies low-amplitude cousins of the Mira variables, which are acknowledged to be pulsating), the number of such stars is small.

If we can demonstrate that there really are no pulsating K-type giants, then a nice *cordon sanitaire* will separate the bottom edge of the pulsating star domain from the top edge of the starspot variable domain, no messy overlap.

6. The New Recipe for Large Starspots

Recently it has been shown (Hall 1991b, 1994) that large starspots are found only on stars in which the Rossby number (ratio of rotation period to convective turnover time) is sufficiently small, smaller than about $Ro = 0.65$. This supposedly is the threshold for strong stellar dynamo action to be triggered.

The dividing line between large and small spots was taken to be those which produce variability (by rotational modulation) larger or smaller than 0.01 magnitude. For comparison, note that the Sun is a weak stellar dynamo and has spots or spot groups which produce variability only 0.001 or 0.002 magnitude in amplitude.

If Rossby number is useful as a single parameter which can segregrate the weak dynamo from the strong dynamo stars, one might reasonably anticipate that all stars with Ro < 0.65 will have large starspots. Note that Hall (1991b, 1994) demonstrated only that no stars with Ro > 0.65 have large starspots, which is only half of the picture.

Thus the finger is pointed at the remnant of Ro < 0.65 stars in which measurable photometric variability, greater than 0.01 mag, has never been demonstrated. Table 5 is a list of 28 such stars, taken from the total of 357 considered by Hall (1994). To allow realistically for uncertainties in the process, the cutoff has been taken at Ro < 0.50. The other half of the picture will have been completed if any of the following proves true. (1) the spectral type or B-V or luminosity class, used to estimate the convective turnover time, turns out to be in error and the resulting correct turnover time gives Ro > 0.65. (2) the rotation period turns out to be in error and the resulting correct value gives Ro > 0.65. (3) additional photometric monitoring finds significant variability, attributable to starspots, afterall.

Since preparing the talk which led to this paper, I have been a co-author on two papers now in press (Crews et al. 1995, Kaye et al. 1995) in which an extensive data base of photometry of eight ellipsoidal variables was analyzed. All eight were previously named variable stars, but the variability had been attributed to the ellipticity effect. Variability due to starspots had been suspected, on grounds of various indicators of chromospheric activity and also the previously mentioned Rossby number test, but never substantiated. All eight *were shown to have starspots* responsible for an additional component of the overall variability, with amplitudes as large as 0.09 mag. Six of those eight (V350 Lac, ζ And, EE UMa, BL CVn, UV CrB, and V826 Her) are among the 28 in Table 5. So, the finger now points at only 22.

7. A Totally New Mechanism for Stellar Variability?

In the process of examining 357 stars and establishing the Rossby number threshold for starspot variability (Hall 1994) I noticed two eclipsing binaries, VZ CVn and PV Pup, which appeared as anomalies in the sense that one component in each was significantly variable (by 0.08 and 0.05 magnitude, respectively) and sufficiently early in spectral type (F0 and A8, respectively) not to be convective and hence not expected to have starspots at all, but not obviously a variable of any previously known type. References to those two eclipsing binaries are Popper (1988) and Vaz & Andersen (1984), respectively. Since then I have learned that a few other astronomers have become aware of a few other stars which are anomalous in the same way: spectral type around F0, measurably variable, but not easy to explain as either pulsating or spotted variables. Preprints of two papers now in print (Krisciunas et al. 1993, Mantegazza, Poretti, Zerbi 1994)

Table 5. Rapid rotators (Ro<1/2) predicted to be spotted variables.

log Ro	ampl. (mag.)	dupl.	lum. class	star name	GCVS name
-1.004	0.000	1	V-IV		RW Aur
-0.660	0.000	1	IV	η Boo	
-0.732	0.000	1	IV		RY Tau
-1.221	0.000	1	III	ψ^3 Psc	
-0.885	0.000	1	III	HR 1023	
-0.821	0.000	1	III	ϵ Hya	
-1.455	0.000	1	III	31 Com	
-0.701	0.000	1	III	7 Boo	
-0.491	0.000	2	V	HD 27691	
-0.830	0.000	2	V	ξ UMa	
-1.164	0.000	2	V	HD 284163	
-0.491	0.000	2	IV		AI Phe
-0.357	0.000	2	III	33 Psc	
-0.573	0.000	2	III	HD 65195	
-0.462	0.000	2	III	HD 108078	
-0.555	0.000	2	III	ϵ UMi	ϵ UMi
-0.578	0.000	2	III		UU Cnc
-0.423	0.000	2	III-II	5 Cet	AP Psc
-0.558	0.000	2	III-II	HR 1970	
-1.294	0.010	2	V	HD 319139	
-0.854	0.010	2	IV-III	HR 8575	V350 Lac
-1.000	0.010	2	III	ζ And	ζ And
-0.402	0.010	2	III	HR 4430	EE UMa
-0.978	0.010	2	III	HD 115781	BL CVn
-0.979	0.010	2	III	HD 136901	UV CrB
-0.449	0.010	2	III-II	HR 6626	V826 Her
-0.823	0.011	1	IV-III	HD 17144	
-0.591	0.012	1	III	10 LMi	

and references contained within them proved useful in adding 8 more stars to the short list started by VZ CVn and PV Pup.

These 10 are listed in Table 6, with columns for the spectral type, the maximum amplitude which has been observed, and the period or periods which have been manifested in the photometry. Some of the references cited have defined the puzzle very nicely. To be understood as δ Scuti-type pulsating variables, the periods are too long (around a day or two instead of one or two hours) and the spectral types are at the extreme edge of the δ Scuti domain. Even though the expected rotational periods are just about right to explain their variability by rotational modulation, to understood them as spotted variables one has to explain how an F0 star can have large spots, because starspots generated by the dynamo process are thought to need a deep convective layer and the outer convection layer is thought to be insignificant in depth until about spectral type F5. With both interpretations (pulsation or starspots) having problems, it has been

Table 6. A new type of variable star?

HD number	HR number	other name	var. star name	spectral type	max. ampl. (mag.)	periods (days)
27290	1338	γ Dor	γ Dor	F1 V	0.1	0.76, 0.73, 1.47
32537	1637	9 Aur	...	F0 V	0.1	2.70, 1.28
96008	...	...	LL Vel	F0 V	0.03	0.309873
62863	3009	2 Pup	PV Pup	A8 V	0.16*	1.92 or 0.48:
110379/80	4825/6	γ Vir	05859	F0 V	0.04*	?
111828	...	...	...	F0 (V)	0.04	1.8 or 4:
117777	...	...	VZ CVn	F0 V	0.05*	approx. 1
164615	...	...	...	F2 V-IV	0.05	0.81
224638	...	...	...	F0 V-IV	0.09	1.43, 1.23
224945	...	...	...	F0 V-IV	0.06	0.35, 0.43

*increased to compensate for light dilution by the companion star

asked if perhaps we are dealing with some physical mechanism for producing variability never before encountered.

Pointing to the convective turnover times computed by Gilliland (1985), I can add some new perspective which might help a little. The turnover time is long enough, to produce small Rossby numbers with reasonable rotation periods, in main-sequence stars later than about spectral type F5. Earlier than F5 the turnover time drops rapidly to something like 10 minutes for a main-sequence star of about 1.5 solar mass. From that minimum, however, the turnover time rises abruptly again, to something like 10 days for a main-sequence star of about 2.0 solar mass. With a rotation period of 1 or 2 days, typical for F0 V stars, and a turnover time of 10 days, one would get a Rossby number of 0.1 or 0.2, comfortably on the right side of the Ro = 0.65 threshold. This consideration promises to have some significance because these curious "new variables" are so sharply concentrated around F0 and apparently sparse in the range between F0 and F5. This new perspective does not, however, definitively answer the question. For one thing, a main-sequence star of 2.0 solar mass is more like a late A star than an F0 star. For another thing, the convective layer at this point is thought to be superficial, so shallow as to be incapable of generating spots large enough to cause the observed variability, the rotationally modulated component of which can be as large as 0.1 magnitude.

To resolve this question of the F0 V variables, we should move on more than one front. (1) Expand the total count on the master list. Every list I have seen in the literature or by private correspondence includes some stars not on other lists and fails to include all stars on other lists. This suggests the master list can be made to grow. (2) Obtain superior photometric coverage of selected stars on the list. The goal would be to define better the nature of the periodicity or periodicities. Is the variability periodic or quasi-periodic? Is there one period or quasi-period for each star or are their multiple periodicities? Does the profile of periodicities change with time and, if so, on what time scale? (3) Determine better or more directly the important physical characteristics such as mass, radius, rotation period, accurate spectral type, and luminosity class. (4) Look more carefully into the question of convection zone and turnover time in this curious region of the early F stars, with the aim of refining the computations and helping us reinterpret the significance of those computations.

Acknowledgments. Travel to the meeting where this paper was presented was supported by N.S.F. research grant HRD 91-04484 and N.A.S.A. research grant NAG 8-1014.

References

Applegate, J. H. 1992, ApJ, 385, 621
Cox, J.P. 1980, Theory of Pulsation (Princeton: Princeton Univ. Press)
Crews, L. J., et al. 1995, AJ, in press
Duerbeck, H. W., & Seitter, W. C. 1982, in Landolt Börnstein Numerical Data and Fundamental Relationships in Science and Technology, Group VI, Volume 2, Subvolume b, section 5.1.
Gilliland, R. L. 1985, ApJ, 299, 286
Glasby, J. S. 1969, Variable Stars (Cambridge: Harvard Univ. Press)
Hall, D. S. 1990, in Active Close Binaries, ed. by C. Ibanoglu (Dordrecht: Kluwer), p. 95
Hall, D. S. 1991a, ApJ, 380, L85
Hall. D. S. 1991b, I.A.U. Colloq. No. 130, 353
Hall, D. S. 1994, Mem. Soc. Astr. Ital., 65, 73
Henry, G. W., Eaton, J. A., Hamer, J., & Hall, D. S. 1995, ApJS, in press
Hoffmeister, C., Richter, G., & Wenzel, W. 1985, Variable Stars (New York: Springer Verlag).
Kaye, A. B., et al. 1995, AJ, in press
Kazarovets, E. V., Samus, N. N., & Goranskij, V. P. 1993, IBVS, No. 3840.
Krisciunas, K., et al. [11 authors] 1993, MNRAS, 263, 781.
Mantegazza, L., Poretti, E., & Zerbi, F. M. 1994, MNRAS, 270, 439
Percy, J. R. 1993, PASP, 105, 1422.
Petit, M. 1982, Variable Stars (New York: John Wiley and Sons)
Popper, D. M. 1988, AJ, 95, 190
Schwarzschild, M. 1975, ApJ, 195, 137
Strohmeier, W. 1972, Variable Stars (New York: Pergammon Press)
Vaz, L.P.R., & Andersen, J. 1984, A&A, 132, 219.

Integrated Telescope and CCD Control on a PC

Norman L. Markworth

Department of Physics and Astronomy, Stephen F. Austin State University, Box 13044, Nacogdoches, TX 75962 USA

Abstract. The 46-cm telescope at the Stephen F. Austin State University Observatory has recently been outfitted with a Photometrics Star 1 CCD camera system for the purpose of CCD photometry. Due to space and environmental considerations, the camera control system has been merged with the telescope control system, yielding an integrated package that can be run on any MSDOS personal computer.

1. Introduction

The Stephen F. Austin State University (SFA) Observatory has enjoyed computer assisted operation of both the 46- and 104-cm telescopes since the early 1980's (Markworth & Rafert 1984). Both telescopes have, until recently, been equipped with single channel photometers as primary instruments. Software has been locally developed (Markworth & Rafert 1987) that has as its primary goal the integrated control of both the telescope and the instrument package. The users of the telescopes include interested undergraduate students as well as graduate students (a terminal Masters' degree is available) and faculty. The diverse backgrounds of the telescope users have driven the software development along the lines of wholly robotic telescopes. When program defaults are selected, the user is primarily responsible for fine centering objects in the field of view and monitoring the quality of the data. Telescope moves are preprogrammed for efficiency and observing sequences are selected for precision results and minimal filter wheel rotation.

Last year the 46-cm telescope was equipped with the Star 1 CCD camera system from Photometrics, Ltd. The computer controlled eight position filter wheel of CompuScope has been mated to the camera to produce a CCD photometer for this telescope. This paper describes the efforts to integrate the camera and filter wheel control into the existing telescope control software.

2. Instrumentation

The Photometrics Star 1 is a stand alone, cooled, integrating CCD camera and controller. The Thomson CSF TH7883 CDA chip has 384×576 pixels, each 23 microns square and 12 bits deep, so that each pixel will register 2^{12} (4096) different brightness levels. The total light sensitive area is 8.83×13.25 mm, which, coupled to the f/9 optics of the telescope, gives a field of view of 8.8×11.2 arcminutes. The filter wheel from CompuScope contains the Johnson B, V, R, I

filters, the Strömgren b, v, and y filters, and a clear filter, all having the same thickness. The spectral response of each filter was convolved with the response of the Thomson chip to match the necessary bandpass.

The original implementation of the CCD photometer used CompuScope software to control the filter wheel through the printer port of an PC/AT386, Photometrics software to control the Star 1 camera system on the same computer, and local software to control the telescope running on a Commodore 64 (Markworth & Rafert 1987). The 46-cm telescope is housed in a 3.8 m ObservaDome with the observing floor 1.5 m above ground level. The space below the observing floor contains the telescope and dome drive electronics and is climate controlled. When the full system is in place the dome is rather full of equipment. Likewise, much of the equipment needs to be moved into the climate controlled area at the end of the night to ameliorate the effects of the East Texas weather. The two computer operation also was inefficient, since one program was needed to turn the filter wheel, then another program was loaded to obtain the exposures. The design of the software required a significant amount of keyboard work, which is difficult during the observations. Furthermore, the camera software did not write a proper FITS header onto the image file for use by the data reduction software.

3. The Integrated System

Work began on recoding the entire system into one package in April, 1994. The 104-cm telescope is already under the control of a single PC/AT286 that controls the telescope, dome, and the three-channel photometer (Markworth 1994). This software was used as a basis for the new 46-cm control software.

3.1. The Camera Link

The camera/computer link is the National Instruments GPIB-PCIIA board, which provides an IEEE-488 SCSI interface. The Photometrics software is a proprietary MS-Windows application and we lacked access to the source code. Photometrics did, however, generously provide the source code to an earlier DOS control program (pcstar) that used the low level National C- functions. Using these calls we could pass commands to the camera controller. As of this writing the integrated system can perform the following functions:

1. Expose - performs a bias command followed by an exposure for the time specified in option 2.

2. Set Exposure Time

3. File Read - reads an image file from the hard drive, extracts the region of interest dimensions from the header, and writes the image to the camera monitor in the same position from which it was stored.

4. File Write - reads the image stored in the camera controller of dimensions given by option 9 into a ram-drive, writes a proper FITS header to the hard drive followed by the image to the same file, and pads the end of the file to conform to FITS format.

5. Clear - clears the chip.

6. Bias - performs a zero second exposure with the shutter closed to map the bias structure of the chip.

7. Dark - performs a bias command followed by an exposure without the shutter open for a time set by option 2.

8. Gain - toggles the camera gain between the two values provided by Photometrics.

9. Set Box - allows the user to specify a region of interest on the camera monitor. The region of interest is highlighted by a box. The default is the full frame.

10. Flat Field - allows the user to specify how many full frame flat field exposures are to be taken. After the telescope is moved to the dome white spot, tracking is disabled, and the flats are automatically processed and stored to disk. At the end of the night another program can be used to form pixel by pixel medians for all of the flats taken.

11. Sequence - allows the user to specify the number of exposures to be taken and stored on the hard drive. No filter moves are done between exposures. A region of interest may be used.

12. Observing Sequence - uses a built in (or user defined) observing sequence to expose the camera and advance the filter wheel. A region of interest may be used. For example, using this command it would be possible to automatically record exposures sequentially for the blue, visual, and red Johnson filters without pausing for user input between exposures.

An automatic file naming scheme has been devised to reduce the amount of keyboard work required of the user at the telescope. In order to make the files more readily usable, a log file is also kept that simply lists the file name and exposure type on that file (i.e., exposure, flat, bias, dark, etc.). The image reduction software is MIRA (Newberry 1993), which can be run on an PC/AT486. The file were chosen to make multiple file processing possible using MIRA.

3.2. The Filter Wheel

The computer controlled filter wheel from CompuScope also came with proprietary software and was controlled through the printer port. Inspection of the interior of the wheel revealed that the wheel is run using a stepping motor and the wheel position is sensed when a paddle moves across an LED/photodiode pair. The 104-cm telescope has seven stepping motors on the three channel photometer and the RA and DEC motors for that telescope are also stepping motors. The software had built in control for stepping motors, using the MetraByte CTM5 counter/timer card as a frequency source, an internal look up table to keep track of the number of motor pulses issued, and a MetraByte PIO96 for output to the stepping motor control circuit. The 46-cm system uses the PIO12 card, which is functionally similar to the PIO96, but provides only 24 lines of parallel I/O rather than the 96 lines of its bigger cousin. The coding of the filter wheel for the PIO12 was straight-forward and frees up the printer port. Since the filter wheel has no outward markings to allow a visual inspection of the wheel position, it is important that the software provide a means of internally sensing the position. The sensing hardware mentioned above gives only one fiducial mark on the wheel. All other moves are relative. In our application the move to the clear filter position is combined with a seek of the sense position. Accurate wheel positioning is thus guaranteed, even in the event of a power loss.

The computer, camera controller, and CCD photometer will be stored in a specially designed cabinet that can be placed below the floor for daytime storage or raised to the observing floor for nighttime use. The cabinet is raised and lowered by counter weights on a cable and pulley system.

4. Observational Program

The system was installed in mid-June 1994 with the first use to be photometry of open clusters. We are specifically interested in discovering new δ Scuti variables. The δ Scuti stars are a populous class of low-amplitude (0001 - 03) short-period (002 - 03) pulsating stars with spectral types A2 through mid-F and luminosity classes V - III. Hundreds of δ Scuti stars have been identified, and there is strong evidence that they exist in much larger numbers at lower amplitudes (Schutt 1991, Breger 1979). Most δ Scuti stars are thought to be Population I although a few have properties associated with Population II (low metallicity, high space velocities, globular cluster members). The pulsation properties of δ Scuti stars, especially the low-amplitude variables, are varied and not well explained by theory. Some δ Scuti stars show single radial modes of pulsation while others have combinations of radial and non-radial modes. Several key questions remain concerning this interesting and important class of variable stars, such as:

1. Do δ Scuti stars of similar age and composition, such as those within a cluster, have similar pulsational properties?

2. What role does stellar evolution play with the observed frequencies and amplitudes? What determines the modes of pulsation?

3. What limits are there to the amplitude of the pulsation?

4. Is there any relation to other pulsating variable stars with similar physical properties (Luminosity, Temperature, Mass, Radius)?

Our study would be the first detailed study of δ Scuti stars in clusters using CCD methods. Research of δ Scuti stars in the past was done star by star and required prohibitive amounts of data. CCD's offer several advantages to this field of research. Gilliland & Brown (1992) demonstrate that low-amplitude δ Scuti stars can be reliably detected in clusters using CCD techniques.

References

Breger, M. 1979, PASP, 91, 5.
Gilliland, R.L. & Brown, T.M. 1992, AJ, 103, 1945.

Markworth, N.L. & Rafert, J.B. 1984 in Microcomputers in Astronomy II, ed. R.M. Genet and K.A. Genet, (Fairborn Observatory), p.9-1.

Markworth, N.L. & Rafert, J.B. 1987 in New Generation Small Telescopes, ed. D.S. Hayes, R.M. Genet, D.R. Genet, (Mesa: Fairborn Press), p. 267.

Markworth, N.L. 1994, "Multichannel Photometry at the SFASU Observatory," AIP Conference Proceedings, in press.

Newberry, M.V. 1993, "Microcomputer Image Reduction and Analysis," 2.0 rev 1, Axiom Research, Inc.

Schutt, R.L. 1991, AJ, 101, 2177.

The Berkeley Automatic Imaging Telescope: An Update

Richard R. Treffers, Alexei V. Filippenko, Schuyler D. Van Dyk, Young Paik

Department of Astronomy, University of California, Berkeley, CA 94720-3411

Michael W. Richmond

Department of Astrophysical Sciences, Princeton University, Princeton, NJ 08544

Abstract. The design and implementation of the Berkeley Automatic Imaging Telescope are presented. The scheduling algorithm is discussed. Applications such as long term photometry of supernovae and the Leuschner Observatory Supernova Search are briefly noted.

1. Introduction

This paper discusses the progress of the Berkeley Automatic Imaging Telescope (BAIT), updating previous discussions (Richmond, Treffers, and Filippenko 1993, and references therein). The 0.5-m telescope began taking data automatically with this software in January 1992; it was joined by the 0.76-m telescope (controlled by the same workstation) in November 1992. The telescopes have been used by many throughout the country for research and instructional projects. One of the pleasures of robotic astronomy is that objects for a wide variety of programs may be observed in a single night.

2. Concepts

The BAIT relies on many features of the Internet. Observers submit request files via electronic mail. A program at the observatory verifies the syntax and legitimacy of these requests and either rejects them or passes them on to the scheduling program. During the night the telescope requests are sorted and scheduled, and the observations are made. The resulting images or photometric points are then transferred to an anonymous ftp site and the observer is notified by e-mail the following morning.

The request file is composed of keyword/value strings and looks like a FITS header. The request file contains a "PROCEDUR." keyword which tells what type of observation is to be performed. This gives great flexibility to the type of observation that can be made. Although the bulk of the observations generate flat-fielded images, other procedures take pictures of planets (using a real time ephemeris), generate bias and/or dark frames, or perform a search for supernovae. In fact a PROCEDUR is simply a call to the operating system so that anything can (in principle) be executed.

3. Implementation

We use a Sun Sparcstation II running UNIX as the master controller. All time-critical functions are done via hardware controllers (e.g., motor controllers, i/o cards) that plug into PCs; we do not depend on quick response from any of the three computer operating systems used at the observatory.

The control program talks to the devices using ASCII messages to a UNIX socket. These messages are relayed to the PCs via ethernet or serial lines. A typical message to position the telescope is "point ra=12:34:33 dec=45:34". All the communication is time-stamped with the Universal time and logged for debugging and postmortem analysis.

The control programs are written in either C or C-shell. Our image processing uses an in-house suite of programs (XVista) written in C for X-Windows. This library is based on the syntax (but not the code) of the Lick Vista program (Stover 1988). The programs use command line arguments (no menus) and thus are easily incorporated in C-shell procedures. We do flat fielding, photometry, and supernova searching using these programs.

We try to avoid "states" in our software. Before each observation, we prepare the observatory to take the observation, making sure that the tracking is on, the slit is open, etc., instead of assuming them to remain that way from a previous observation. While this procedure takes a few seconds longer, it makes the system more tolerant of any failures that may occur.

We make use of the FITS header in the data file to log information and transmit it to other pipelined programs, rather than creating second files that may lose this information. The FITS header is also generated by reading the actual state of the telescope and camera rather than relying on what commands were sent. This somewhat cumbersome procedure has prevented problems in a multi-tasking environment where other programs may be setting values of filter positions, etc.

4. List Generation

Our list generator algorithm attempts to schedule each request as close to transit as possible in the following manner.

1. Each request file is assigned a numerical rank which depends on a user assigned priority and on whether the observation is overdue:

$$rank = Priority + f(days\ since\ last\ interval).$$

The function f has the property that it demotes observations that are not yet due to be observed. If two requests of equal priority are both past due, then the most recently observed one is demoted.

2. Sort all the requests in rank, putting the most important first (lowest numbers in our scheme).

3. Calculate the observing window for each request taking into account rise time, twilight, moon position/phase, and any other special user requirements. An estimate of the observation time (exposure, telescope move time plus overhead) is used. (The order of the objects scheduled in the early part of the list is quite insensitive to the observation time estimates.) The time of best observation (nearest meridian if possible) is calculated.

4. Start the list by placing the highest ranked object at its best time.

5. Place subsequent requests at their best times, or as near to their best times as possible.

6. If an observation cannot fit, then push the ones already scheduled aside (yet keeping their order) and see if it can be jammed in. Observations, once scheduled, cannot be pushed out of their allowed windows. Early in the project, we were concerned that this procedure would move the highest priority observations far from their optimum positions, but in practice this has not occurred since the scheduled observations have so many constraints that their motion is limited. The pushing merely "wrings out" the gaps in the schedule.

7. If an observation can't fit, give up on it and go on to the next one (starting at step 5 above) until all of them have been tried.

8. Finally, when the list has been prepared, simply pick the first object scheduled and observe it, discarding the rest of the list. It takes us about 8 seconds on a Sparcstation II to schedule 200 requests.

9. Often you may have to wait (a few minutes) before starting the next observation since it is not yet the optimum time although it is still within its legal observing window. When this occurs we check whether the list is overbooked at this time (by seeing if there were rejected requests for this interval); if it is, we start the next observation immediately. This procedure wrings out small unneeded dwell periods.

We have been quite pleased with this algorithm because it schedules observations for scientifically important reasons (low airmass) rather than schemes that try to minimize telescope motion or other factors. Since the list is regenerated after each observation, it never gets out of date. If an observation fails or takes longer than expected, no great loss occurs. Also, since the list is regenerated after every observation, adding new requests during the night ("dynamic scheduling") is simple.

In our supernova search, we "cluster" our observations in groups of galaxies within a few degrees of each other, generating requests that take about 5 to 10 minutes to observe. This way we cut down on the number of targets to schedule and minimize telescope and dome motion.

5. Scientific Programs

The BAIT project currently uses the 0.5-m and 0.76-m telescopes at the Leuschner Observatory located less than 10 miles east of the Berkeley campus. The site has less than ideal seeing, but is convenient for development. The 0.5-m telescope is an $f/11.5$ classical Cassegrain. The camera uses a Photometrics Ltd. Peltier cooled camera with a Metachrome coated Thomson 512×512 pixel CCD. The autoguider and method used for automatic guide star acquisition are described in Treffers, Richmond, and Filippenko (1992; see also Richmond, Treffers, and

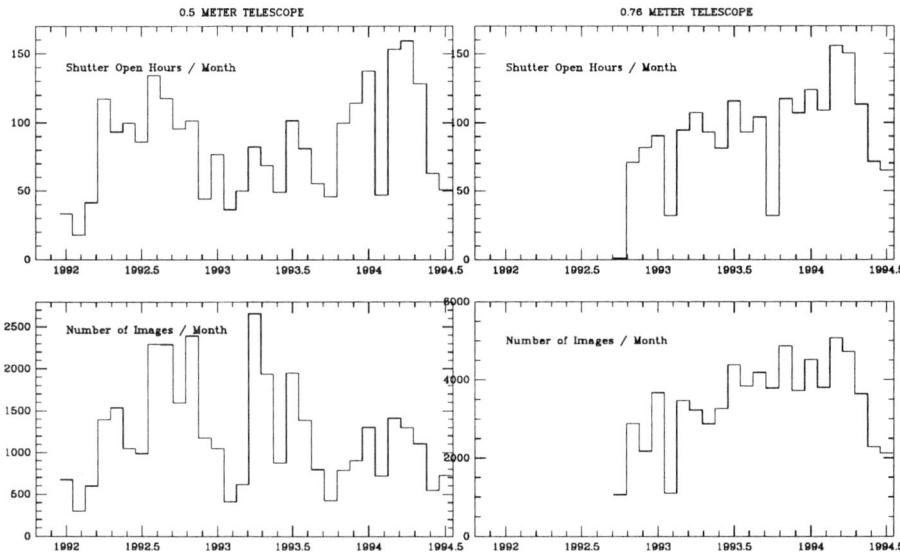

Figure 1. Leuschner Telescope Usage.

Filippenko 1992). The 0.76-m telescope is an $f/8$ Ritchey-Chrétien. It has a Photometrics Ltd. Peltier camera with a 516 × 516 pixel Metachrome Ford Aerospace CCD (on loan from Lawrence Berkeley Laboratory). Figure 1 shows the telescope usage.

Since the 0.5-m telescope has an autoguider, we use it to study fainter objects, mostly pursuing long-term photometry of supernovae and active galactic nuclei (see also Filippenko 1992). We have published studies of the subluminous Type Ia supernova 1991bg (Filippenko et al. 1992), three other Type Ia supernovae (Ford et al. 1993), and the bright, peculiar Type II supernova 1993J in M81 (Richmond et al. 1994). Preliminary photometry of the recent Type Ia SN 1994D (Richmond et al. 1995) is shown in Figure 2.

The 0.76-m telescope plays a role in many projects. University of California (Berkeley) Astronomy courses use it, both for visual observing and remotely when learning observational techniques. The Lawrence Berkeley Laboratory (LBL) "Hands-On Universe Program" (Barclay et al. 1992) allows high-school students to request their own astronomical images. In the Spring of 1994 two students (Heather Tartara and Melody Spence) from a high school in Oil City, Pennsylvania used it to obtain the earliest known image of SN 1994I in M51.

In the Winter of 1992 we began a search for supernovae in nearby galaxies visible from the northern hemisphere, with particular emphasis on early detection. The software and target list differ from those of the search previously run by LBL at this telescope (Muller et al. 1992). A brief discussion can be found in Treffers et al. (1993). So far we have discovered six supernovae; a summary of the control time is shown in Figure 3.

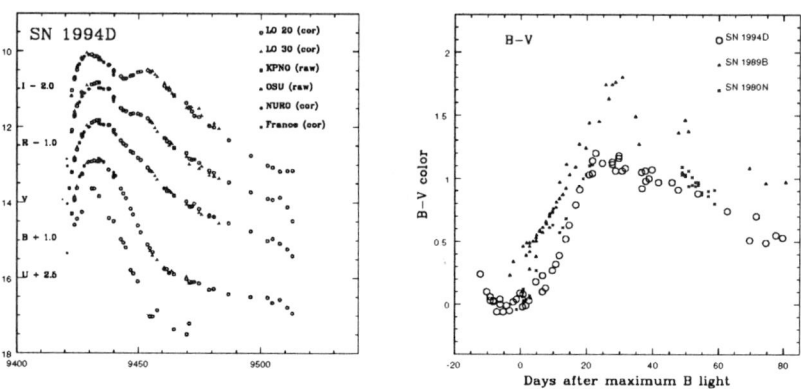

Figure 2. SN 1994D Photometry.

6. Move To Lick Observatory

We are currently moving the camera and guider from the 0.5-m telescope to a new 0.76-m robotic telescope in the former 0.6-m telescope dome at Lick Observatory. We will be using an AutoScope telescope controlled by the BAIT software. We plan on taking advantage of the improved seeing and darker skies afforded by this site.

Acknowledgments. We acknowledge the equipment donations or financial support of AutoScope Corporation, Photometrics Ltd., Sun Microsystems, the Sylvia and Jim Katzman Foundation, and the National Science Foundation.

References

Barclay, T., Lenk, C., Arsem, E., Desai, J., Monsen, G., Govil, K., and Pennypacker, C. 1992, in Hands-On Astronomy for Education, ed. C. Pennypacker (Singapore: World Scientific), p. 63.
Filippenko, A. V. 1992, in Robotic Telescopes in the 1990s, ed. A. V. Filippenko (San Francisco: ASP), ASP Conf. Ser. 34, p. 55.
Filippenko, A. V., et al. 1992, AJ, 104, 1543.
Ford, C. H., et al. 1993, AJ, 106, 1101.
Muller, R. A., Newberg, H. J. M., Pennypacker, C. R., Perlmutter, S., Sasseen, T. P., and Smith, C. K. 1992, ApJ, 384, L9.
Richmond, M. W., Treffers, R. R., and Filippenko, A. V. 1992, in Robotic Telescopes in the 1990s, ed. A. V. Filippenko (San Francisco: ASP), ASP Conf. Ser. 34, p. 105.
Richmond, M. W., Treffers, R. R., and Filippenko, A. V. 1993, PASP, 105, 1164.
Richmond, M. W., Treffers, R. R., Filippenko, A. V., Paik, Y., Leibundgut, B., Schulman, E., and Cox. C. V. 1994, AJ, 107, 1022.
Richmond, M. W., et al. 1995, in preparation.

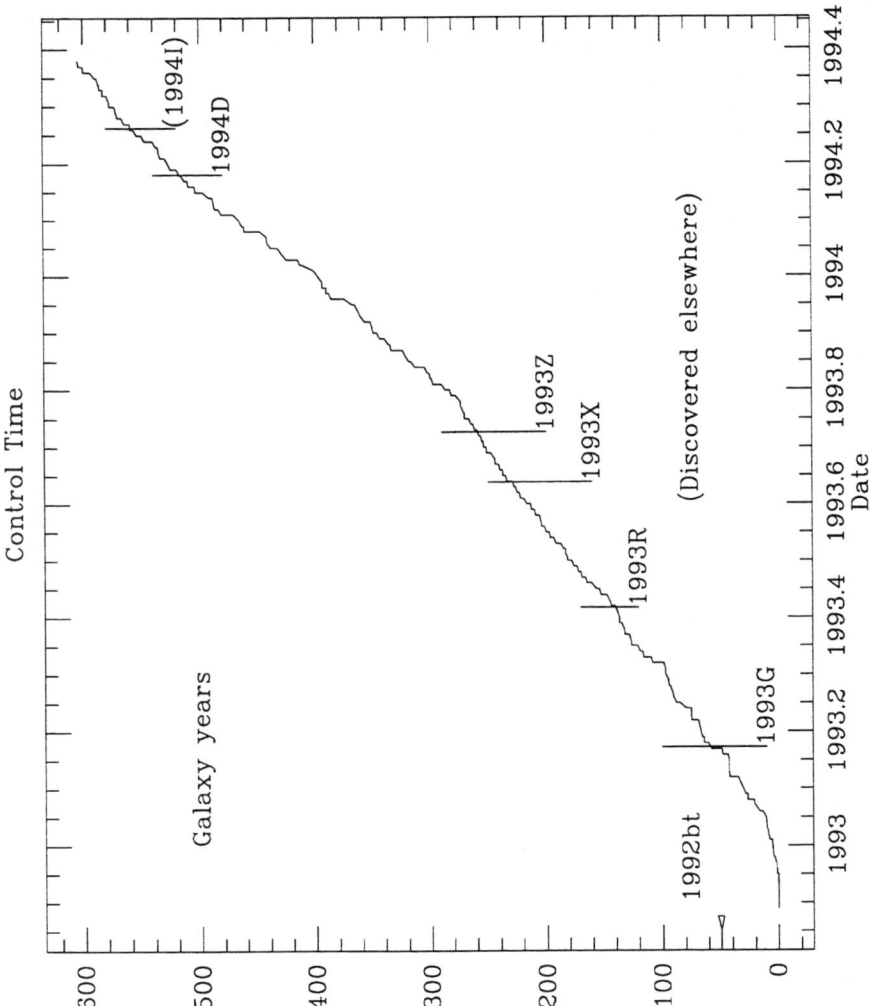

Figure 3. Leuschner Observatory Supernova Search.

Stover, R. J. 1988, in Instrumentation for Ground-Based Optical Astronomy, ed. L. B. Robinson (New York: Springer-Verlag), p. 443.

Treffers, R. R., Richmond, M. W., and Filippenko, A. V. 1992, in Robotic Telescopes in the 1990s, ed. A. V. Filippenko (San Francisco: ASP), ASP Conf. Ser. 34, p. 115.

Treffers, R. R., Leibundgut, B., Filippenko, A. V., and Richmond, M. W. 1993, BAAS, 25, 834.

THE MICRO-OBSERVATORY: AN AUTOMATED TELESCOPE FOR EDUCATION

P. STEVEN LEIKER, and PHILIP M. SADLER Harvard-Smithsonian Center for Astrophysics, 60 Garden Street, Cambridge, MA 02138

KENNETH BRECHER, Departments of Astronomy and Physics, Boston University, Boston, MA 02215

ABSTRACT For the past several years, a team of scientists, engineers, and teachers has been developing a portable, low-cost, automated imaging telescope designed specifically for classroom use which can be used in real-time or batch mode. The initial testing of the MicroObservatory has been completed. Students have used the telescope to observe celestial objects and have brought out detail and faint features in captured images with easy-to-use public domain image-processing software. In addition to the telescope control and image processing software, we are developing pattern recognition software that will allow the computer to aid the user in field and object identification. We are now in the process of building five second generation MicroObservatory telescopes, which will be tested in schools. Teachers have expressed great enthusiasm for the large-scale implementation of the project.

INTRODUCTION

With an ordinary optical telescope, the difficulty and specialized knowledge required to set up, point, track, photograph, and interpret astronomical observations are beyond the capability of most students and high school science teachers. Compounding these problems is the difficulty of getting students safely and comfortably to an observing site in coordination with weather appropriate for observations. The MicroObservatory solves these problems and provides two further benefits: 1) the images taken with the MicroObservatory can easily be processed with the MicroObservatory host computer or other computer, permitting abundant individual experimentation, and 2) the advanced capabilities of the MicroObservatory telescope permit students, in coordination with other amateur and professional astronomers and associations (such as the American Association of Variable Star Observers) to participate in real astronomical research. The MicroObservatory is a tool for original nighttime observations, as well as for daytime observations of the Sun, Moon, and possibly other bright objects such as Venus. The MicroObservatory Project was first funded by the National Science Foundation (NSF) in 1989, which resulted in the development of the first prototype MicroObservatory. The project was funded by the NSF for a second time in 1992 to produce five second generation telescopes to be tested in high schools.

THE FIRST MICRO-OBSERVATORY PROTOTYPE

The MicroObservatory is a totally computerized telescope controlled by a Macintosh II series computer (Buil, Sadler, Brecher 1994). Traditionally, one of the most difficult aspects of observational astronomy is setting up and using a telescope. Many amateur and professional astronomers spend a great deal of time assembling, troubleshooting, and calibrating the equipment. The MicroObservatory will allow any person to immediately start making astronomical observations without encountering many of the problems of the traditional observatory. This level of automation is accomplished by using two computers: one computer is built into the MicroObservatory, and the second computer is the Macintosh. All of the functions of the MicroObservatory telescope are controlled by the built-in, or on-board, computer. This on-board computer receives commands from the Macintosh and then appropriately controls the telescope. The MicroObservatory is capable of astronomical observations which can be used not only for pedagogical purposes but also for astronomical research.

The MicroObservatory is a portable, self-contained telescope. The height of the telescope is about 1 meter and it weights approximately 36 kg. For easy transportation it disassembles into two pieces of about 18 kg each. Figure 1 is a diagram of the first generation MicroObservatory. The MicroObservatory has an aperture of 15 cm and contains an astronomical-quality CCD camera. The telescope is connected to a Macintosh computer through a 57.6K baud serial port.

The height of the telescope was kept to a minimum since people do not have to look through it as they would with a traditional telescope. The low height of the MicroObservatory is an advantage because it allows the structure of the telescope to be very rigid. It also minimizes wind induced movement. The MicroObservatory telescope has the conventional (though computerized) right ascension and declination axes like other equatorial telescopes, but in addition, it has computerized altitude and azimuth axes. The advantage of having the altitude and azimuth axes computerized is that the polar alignment procedure can be automatic, making it possible for almost anyone to accomplish rapidly. The only manual part of the telescope polar alignment procedure is the leveling of the telescope base. Each axis of the telescope is driven by a stepper motor which is controlled by the telescope's on-board computer. A cable connects the MicroObservatory to its power supply. This cable also provides communication lines to the telescope. The Macintosh computer connects to the power supply with a normal serial communication line.

The optical design of the first generation MicroObservatory telescope is a 15 cm folded-Newtonian. A zoom-type camera lens is used to increase or decrease the effective focal length of the optical system. The focal length range is from 850 mm to 2530 mm. The effective focal length of the telescope is jointly controlled by the on-board computer and the Macintosh. The user can easily instruct the computer to change the effective focal length of the telescope whenever the user desires to do so. In a more automated mode, the computer selects the proper focal length. Images are obtained with a cooled CCD camera. The CCD imager is thermoelectrically cooled to approximately 40°C below ambient temperature. After an image is obtained, it is transferred to the Macintosh computer for storage and image enhancement. The optical assembly also contains a filter wheel which allows the user to select one of several different filters through which to image. In a more automated mode, the computer selects the proper filter.

THE SECOND GENERATION MICRO-OBSERVATORY

We are now in the process of building five more MicroObservatory telescopes, which will be tested in the field. We are making many improvements to the original design. The following list highlights some of the improved features of the new instruments:

- Higher spatial resolution CCD imager;

- Telescope to Macintosh communication at Ethernet speeds to cut down on the time it takes to transfer an image;

- Improved microstepper driver for the right ascension drive;

- Higher telescope slew rates in right ascension and declination;

- A Maksutov optical system;

- Improved software to make the telescope more scientifically useful and easier to operate; and

- Remote operation of the MicroObservatory over the Internet.

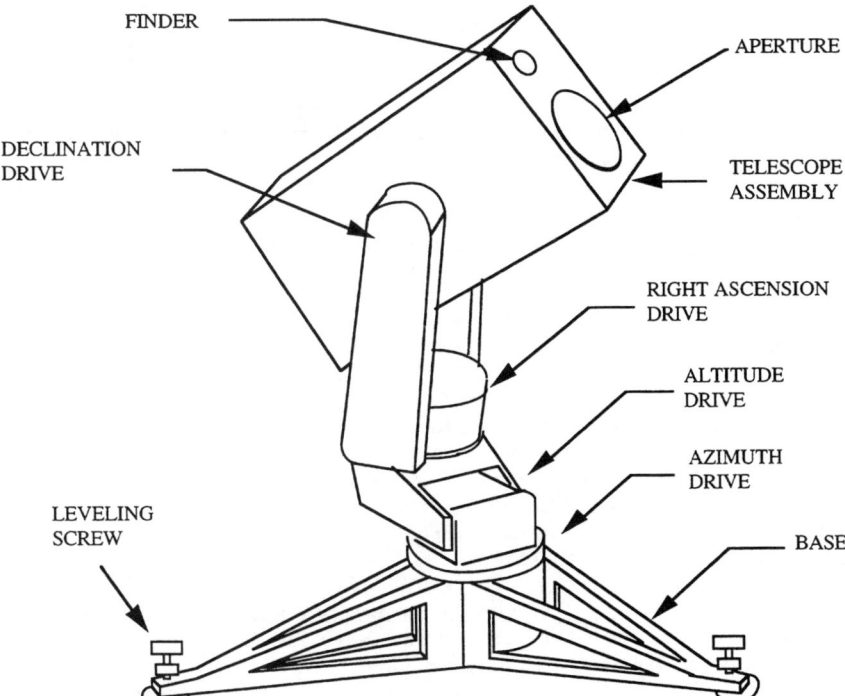

Fig. 1. The MicroObservatory

Over the next year, we will complete five telescopes and develop a network of these portable instruments that will allow students to contribute to current research in astronomy. Research astronomers and active amateurs will oversee observational programs for which the MicroObservatory is particularly suited; teachers will be trained in its use; and a system will be implemented whereby students can collect data, use the Internet to transmit the data to a central research database, and share in the analysis of data collected from other sites.

MICRO-OBSERVATORY SOFTWARE

The software takes advantage of the Macintosh user interface, making the MicroObservatory powerful, yet very easy to use. From the Macintosh computer, the user can direct the telescope, select an object for observation, take an image of the object, and process the image with NIH (National Institute of Health) Image, a public domain image-processing program or other image processing software. An object may be selected by name, by coordinates, or by an ephemeris. The control software allows the user to choose resolution, exposure times, filters, and other telescope and camera parameters, or to use an automatic setting for these parameters. As with most other research instruments, the only way to view an object through the MicroObservatory is to look at an image on a computer screen. Image-processing software allows the user to manipulate an image and discover more information from the image than can be seen initially.

The software operates in two modes: real-time and batch mode. In the real-time mode the coordinates for an object can be directly entered into the control program, or a planetarium program can be used to point the telescope. Using the planetarium program one can view the sky from the selected location and time. By moving the cursor to any object that is displayed on the screen, the program will display descriptive information about that object. At the press of a button, the telescope will move to the selected location. The student can then take an image of that object and decide on other options such as changing the focal length, or type of filter. Once an image is obtained by the MicroObservatory, it can be stored for future access. In the batch mode, students can select objects ahead of time and the images will automatically be obtained and stored.

The software that drives the MicroObservatory can be broken into three parts: 1) in-telescope software, 2) communication and control software, and 3) the "front-end" or the user interface. The control software runs in the Macintosh and is written in Think C. The control software is the link between the user interface and the in-telescope software. The user interface is easy to understand and is very "Macintosh-like", operating very much like most other Macintosh applications. The planetarium program that is used is a special version of the popular Voyager planetarium program. Tom Mathis of Carina Software adapted a version of Voyager (which he developed) so that it could be incorporated into our MicroObservatory control program.

In addition to the telescope control and image processing software, we are developing pattern recognition software to aid the user in identifying stars and star fields. This software was based on the work of Groth (1986). We have created a subset of the Hubble Guide Star Catalog that forms the basis of our on-line catalog. The on-line catalog contains all stars to 13th magnitude, and the names and numbers (HD, SAO, etc.) of all stars to the 8th magnitude. The pattern matching program matches a MicroObservatory image with stars from that area of the sky from the on-line catalog. The amount of time required to find a match varies with the number of stars in the field, but averages approximately 20 seconds with an older

(Mac IIvx) computer. Once an image is matched, an overlay is drawn on the computer screen with the name and number of each star in the field, for stars with a name or a number in the database.

ASTRONOMY EDUCATION

Probably in no other area of science is it possible for a broad spectrum of high school students (not just the gifted) to have access to equipment which would allow them to do original research. We see several important additions to the educational enterprise through the development and broad dissemination of the MicroObservatory:

- For the large number of earth science, physical science, physics, and astronomy courses taught in the schools, this is a system which would allow for astronomical observations in the daytime when students are in school, or batch mode for night observations when the students are at home, making astronomy a laboratory science like others in the schools;

- Students themselves could undertake a wide range of research projects which would allow them to learn by doing rather than just by reading, listening to lectures or performing standard laboratory exercises; and

- Students would develop a variety of skills, including computer programming, image processing, data analysis, as well as planning and designing their own original projects.

FUTURE PLANS

We believe the MicroObservatory will be a very effective teaching tool. It will give students of a wide range of ages, abilities, and backgrounds who normally would not have access to a telescope the opportunity to use a powerful scientific instrument, along with the possibility of doing original research. This project supports the formation of a network of these portable instruments which will allow students to contribute to current research in astronomy. To test this idea, we are organizing a group of research astronomers to oversee observational programs for which the MicroObservatory is particularly suited. We will train teachers in its use and implement a system whereby students can collect data, use a computer network to transmit data to a central research database, and share in the analysis of data collected from many sites. To accomplish this, we plan to:

- finish construction of five MicroObservatory telescopes and deploy them throughout the U.S.;

- organize a cadre of teachers to act as agents for testing the MicroObservatory;

- design and administer a telecommunications network for research collaboration; and

- organize a group of professional astronomers to work with our student network.

This project will help to investigate the role of student-scientist collaboration--the research "apprenticeship". Students will also get valuable experience with the delicate combination of competition and cooperation that permeates the scientific enterprise.

REFERENCES

Buil, C., Sadler, P., and Brecher, K. 1994 Advanced Imaging, 9, 41.
Groth, E. J. 1986 AJ, 91, 1244.

PART II

PRESENT DEVELOPMENTS

Flexible Scheduling of Automatic Telescopes over the Internet

Mark Drummond
Recom Technologies
NASA Ames, MS: 269-2
Moffett Field, CA 94035-1000

John Bresina
Recom Technologies
NASA Ames, MS: 269-2
Moffett Field, CA 94035-1000

Will Edgington
Recom Technologies
NASA Ames, MS: 269-2
Moffett Field, CA 94035-1000

Keith Swanson
Computational Sciences Division
NASA Ames, MS: 269-2
Moffett Field, CA 94035-1000

Greg Henry
Tennessee State University
330 Tenth Avenue North
Nashville, TN 37203-3401

Ellen Drascher
Recom Technologies
NASA Ames, MS: 269-2
Moffett Field, CA 94035-1000

Web page: `http://ic-www.arc.nasa.gov/ic/projects/xfr/`

Abstract. This paper outlines a new telescope operations paradigm that is intended to achieve low operating costs with high operating efficiency and high scientific productivity. The paradigm is based on the existing *Principal Astronomer* approach used in conjunction with ATIS, a language for commanding remotely located automatic telescopes. This paper describes a telescope management system called the *Associate Principal Astronomer*, or APA. At the heart of the APA is automatic observation loading and scheduling software, and it is this software that is expected to help achieve efficient and productive telescope operations. The purpose of the APA is to make it possible for many astronomers to submit observation requests to, and obtain resulting astronomical data from, a remotely located automatic telescope. All communication is conducted via the Internet in a highly-automated way that minimizes day-to-day human management of the system and that maximizes the scientific return from observing time.

1. Introduction

This paper presents an instance of a new paradigm for telescope operations. The general paradigm is based on the idea of specifying an observing request in a programming-language-like fashion and sending this specification to a telescope that is able to automatically execute it. The instance of this paradigm discussed in this paper is based on one particular language for specifying observing requests: the Automatic Telescope Instruction Set, or ATIS(Boyd, *et al.*, 1993). ATIS has recently undergone formalization by the International Astronomical Union, and the result is ATIS93, a language well-suited for specifying both observing requests, and, interestingly, the order in which those requests should be executed. This paper discusses ATIS-based telescope operations and presents a general design for a system that works with ATIS telescopes. The system is called the *Associate Principal Astronomer*, and it is designed to support multiple astronomers accessing a remotely located, fully automatic telescope, over the Internet.

2. Background

Research-quality telescopes located at prime observing sites have always been a scarce resource, and astronomers have typically had to work with extremely limited access. With manual observations, observing time is allocated to an individual astronomer a few times per year in short contiguous blocks of a few nights each. Furthermore, an astronomer has needed to be physically present at a given telescope in order to operate his or her instrumentation for data acquisition. Limited access, block allocation, and local operation have restricted both the amount of data that can be gathered and the type of observational campaigns that can be accomplished. In recent years, two new observing paradigms have been developed to overcome the problems associated with this traditional telescope operations model.

2.1. Teleoperation

Electronic network and communication technologies have enabled a new observing paradigm where astronomers may participate in an observation program from a remote location. Various approaches have been considered, from remote verbal communications with the on-site telescope operations staff, to actual remote control of a telescope with real time video feedback (see Emerson and Clowes, 1993). Such approaches to remote observation provide flexibility by allowing the observer to be physically distant from the telescope yet to remain in direct control. However, even with this remote observing paradigm, an astronomer must still be involved during the execution of an observing program, and human presence at the observatory is often still required. We refer to this paradigm of remote observation as *teleoperation*, since an astronomer is in direct control of the telescope while his or her observational campaign is performed.

There are a number of problems with teleoperation as a telescope operations paradigm. Primarily, a telescope is not used *efficiently*, and this is a result of a number of factors. First, while teleoperating a telescope, an astronomer repeatedly decides which object to observe next. When making such a decision, the telescope is often idle, waiting for a command from the astronomer.

A second inefficiency in teleoperation results from *block allocation* of observing time. Block allocation typically means that an astronomer is given a number of consecutive nights of observing time on a given telescope. With fine-grain block allocation, an astronomer is given a number of consecutive hours of observing time on some particular night. Allocating a block of time is the only way that teleoperation can work effectively, since it is unrealistic to expect each astronomer to establish a long-distance connection to the telescope, make a single observation, and then disconnect. The overhead of such single-observation teleoperation sessions is simply too great.

Whether allocation of time is in terms of blocks of nights or blocks of hours, the result is similar: during the time that a given astronomer has control of the telescope, no other astronomer's observations can be made. This decreases the operating efficiency of the telescope, since the particular block of time assigned to a given astronomer might not provide observing conditions suitable for that astronomer's campaign. For instance, if the astronomer is planning on doing high precision photometry, and if the hours for which the astronomer is given the telescope have extremely thin cirrus cloud cover, then those hours are wasted, since high precision photometry is impossible. However, to a spectroscopist, those hours are not without merit. Thin cirrus clouds simply result in longer integration times for spectroscopic observations.

Finally, another problem with teleoperation as a paradigm for remote observing is that it can be extraordinarily tedious for an astronomer to carry out repetitive observing sequences. Repetition can occur within a single night, as it is often necessary to observe the same star under a number of different conditions (at different airmasses and at different times of the night). Repetition can also occur over a number of nights, as it is often necessary to observe a star for months, or years, in order to obtain a meaningful account of the star's long-term behavior. For such observational campaigns, teleoperation is an inappropriate and inefficient way to use an astronomer's valuable time.

Rather than require an astronomer to be in remote control of the telescope whenever data is acquired, it is possible to allow the astronomer to specify his or her observing campaign in a domain-specific programming language. This idea gives rise to a rather different remote observing paradigm, that of fully automatic execution.

2.2. Fully automatic execution

Fully automatic telescopes represent an alternative to teleoperation, allowing an astronomer to be removed from the telescope in both space and in time. For example, Fairborn Observatory (Mt. Hopkins, Arizona) has designed and built software and hardware systems for the control of modest-aperture telescopes equipped with photoelectric photometers. These systems make it possible for a remotely located telescope to operate unattended for significant periods (up to a number of months). These telescopes execute commands provided by an astronomer in such a way that the astronomer is not required to participate in the execution of the observing program. It is in this sense that these telescopes are *fully automatic*. (Genet and Hayes, 1989, describe automatic photoelectric telescopes in some detail.) While the majority of existing ground-based automatic telescopes are used for aperture photometry, automation support for spectroscopy and imaging has been increasing (primarily due to the efforts of R. Kent Honeycutt and Don Epand; see Boyd, *et al.*, 1993).

This paper is concerned only with the fully automatic telescope operations paradigm; however, before we continue, it seems reasonable to consider the conditions under which teleoperation is more appropriate than automatic execution. Suppose, for instance, that an astronomer's observing campaign cannot be algorithmically expressed, that is, written down as a program. Clearly, this astronomer must be in direct control of the telescope at the time of observing. Additionally, if observations to be made depend heavily on what has just been observed, and if there is significant human judgment required to determine the particular observations to make, then teleoperation is a reasonable mode of operation. However, when the observing program *can* be algorithmically expressed, and when the sequence of observations is not decided on-the-fly by human judgment, then fully automatic telescope operation is more appropriate and significantly more efficient than teleoperation.

3. An example of fully automatic execution

In this section we first introduce the Automatic Telescope Instruction Set and then discuss the sort of fully automatic telescope operations that it supports. We also briefly evaluate the performance of some telescopes that use this language in order to better define the problems that our software system is intended to solve.

3.1. The Automatic Telescope Instruction Set, ATIS

For the automatic telescopes we are considering, the language used to define observation requests is the Automatic Telescope Instruction Set, or ATIS (Boyd, *et al.*, 1993). In ATIS, a *group* is the primitive unit to be scheduled and executed. A group contains commands to move the telescope, to control the filters, and to gather data in a defined sequence. In the initial version, ATIS89, the only instruments accommodated were photometers, but the most recent version, ATIS93, also includes commands to obtain CCD camera images.

In addition to specifying the syntax and semantics for observation requests and results, the ATIS standard provides a set of *group selection rules* that are used to determine the execution order of groups during the night (see Genet and Hayes, 1989). The group selection rules provided by ATIS essentially implement a first-to-set-in-the-west policy: at any given point in time the telescope observes the star that will set next. Other group selection rules provided by the ATIS standard deal with factors such as priority and execution count, but the first-to-set-in-the-west rule typically dominates group selection.

3.2. How ATIS-based telescopes work

The left half of figure 1 shows the overall flow of requests and results in an ATIS-based telescope operations system.

First, an astronomer forms a set of groups to achieve the scientific goals of his or her observation campaign. For any given automatic telescope, there is

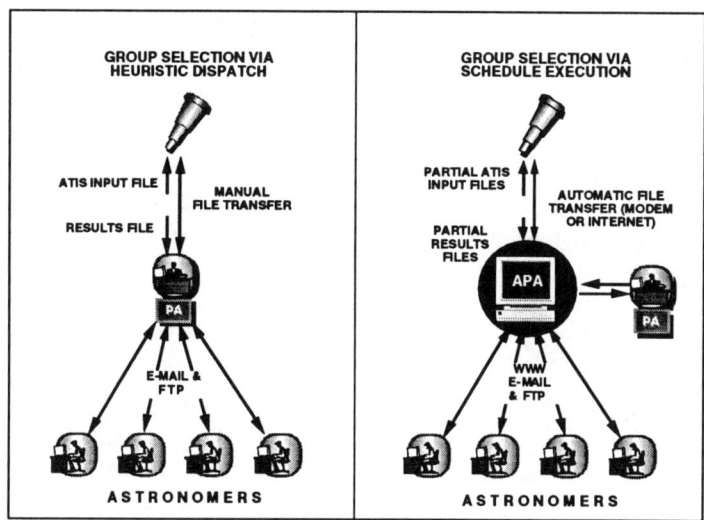

Figure 1.   Life, before (left) and after (right) the APA.

a single *Principal Astronomer* or PA. The PA manages the set of requests that are loaded onto the telescope. Thus, once an astronomer has assembled a set of ATIS groups, these are sent to the appropriate PA, typically via electronic mail (e-mail) or by the Internet's File Transfer Protocol (FTP).

The PA collects together requests from participating astronomers and attempts to ensure that the total set of groups is reasonable – that the telescope load makes good utilization of observing time and is fair to all participating astronomers, that there are appropriate groups for quality control and data reduction, and so on. Then the complete set of groups is sent to the computer that controls the telescope. Communication between the PA and the telescope control computer is typically carried out using personal computers connected via the Internet or over modems and phone lines. The important aspect of this communication is that the PA can be located anywhere on the planet (in principle) and need only have access to some sort of communication link.

The telescope controller uses its built-in ATIS group selection rules to implement a form of heuristic dispatch scheduling (see Genet and Hayes, 1989). At any point in time, the rules recommend a single group to execute next. The groups are executed by the telescope controller for some number of nights (often months); eventually, the PA requests from the controller the results that have been collected thus far. The collected data are returned to the PA as an ATIS results file (the format of which is also specified by the ATIS standard). The results include the raw data obtained from the observations, as well as a chronological record of the groups that were executed and relevant observing parameters to help with data reduction. The PA edits this results file and sends each astronomer the pieces corresponding to his or her requested observations (again, typically via e-mail or FTP). In some cases the PA provides a basic data-reduction service, returning reduced results, not just raw data.

3.3. ATIS-based telescopes can work extremely well

One of us (GWH) has been working as a PA for a number of years with automatic telescopes. Together with Louis Boyd (of Fairborn Observatory) acting as principal telescope engineer, several telescopes have been operated automatically on Mt. Hopkins in southern Arizona to accomplish a wide variety of different scientific programs.

The efficiency of operations for these telescopes has been estimated to achieve a dollar-cost-per-observation that is 30 to 40 times cheaper than previously possible using traditional manual telescope operations (see Henry, "The Fairborn/TSU Robotic Telescope Operations Model", this volume). There has also been an enormous increase in observational efficiency: the combined *yearly* output of the automatic telescopes managed by GWH would require a *lifetime* of effort to obtain by previous manual methods of operation.

Cost and efficiency are not the whole story, of course. Of paramount importance is the precision with which measurements can be made. Henry (see "The Development of Precision Robotic Photometry", this volume) has demonstrated an order of magnitude improvement in automatic telescope photometry, and measurements of millimagitude precision are now routine. This is a dramatic improvement when compared to previous manual photometric observations.

Another benefit of the ATIS operations paradigm is the ability to pursue extremely long-term observing campaigns. Some of the Fairborn telescopes, mentioned above, have been used in support of campaigns lasting many years. In differential photometry, it is common to attempt to measure the light curve for a given variable star. When a variable star has very long periodicities, obtaining sufficient data can be very difficult. ATIS-based fully automatic telescopes directly address this long-term data acquisition problem. For instance, Fairborn telescopes have been used to obtain data on chromospherically active stars like λ Andromedae for the last 12 years (see Henry, "The Development of Precision Robotic Photometry", this volume, figure 13).

3.4. But there are problems with ATIS operations

There *are* some problems with the current operation of ATIS-based telescopes. Most of these problems relate to the way in which the ATIS group selection rules order astronomers' groups for execution at the telescope.

First and probably most important, the group selection rules can easily give rise to a horizon-to-horizon "thrashing" behavior. The ATIS group selection rules start observing stars in the west. The rules specify that the star that is about to set next be observed next (all other things being equal). Under the guidance of these rules, the telescope tends to work its way through the given set of groups, moving from the telescope's western observing limit towards the meridian. The telescope then proceeds towards its eastern observing limit, making measurements of stars as they move towards the meridian. If the telescope reaches the eastern limit before dawn and if the PA has requested multiple observations of groups in anticipation that the telescope might be underloaded, then the ATIS group selection rules instruct the telescope to return to the western observing limit to make another observation of the next star to set. During this observation, however, new stars can have risen in the east, some of which are defined as targets by existing ATIS groups. The group selection rules instruct the telescope to move to the east in order to observe these newly risen stars. Once they have been observed, however, the group selection rules instruct the telescope to return again to the western observing limit to observe the next star to set. This limit-to-limit thrashing behavior can continue for the remainder of the night.

This thrashing behavior has the unfortunate result of making observations through very high airmass. Ideally, and all other things being equal, observations would be made only near the meridian, as it is there that the airmass is minimal for a star of a given declination. Of course, all other things are rarely equal, and it is often necessary to make observations well off the meridian. However, with better scheduling, it is possible to prevent this sort of limit-to-limit thrashing behavior (Drummond, Bresina, and Swanson, 1994).

The thrashing is essentially a result of under-loading the telescope with respect to the default ATIS group selection rules. The entire concept of under-loading is only defined with respect to a particular scheduling policy, in this case, the default ATIS rules. It is also possible to overload a telescope. Using the ATIS group selection rules, an overloaded telescope starts at the western observing limit, but never makes it all the way across the sky to the eastern limit; indeed, with sufficient overloading, the telescope might never make it as far as the meridian. It is important to realize that these sorts of under-loading and overloading behaviors are defined only with respect to a given scheduling policy. The behavior of the telescope is a function of the groups that it is given *and* the policy by which those groups are scheduled for execution. With a different scheduling policy, but with the same set of groups, entirely different execution behavior can be obtained.

Unfortunately, with ATIS89, there is no way to change the scheduling policy "on the fly". The rules by which the scheduling policy are implemented are fixed in ATIS89, so the only variable that can be controlled by a PA is the set of groups sent to the telescope. Thus, to correctly load an ATIS89 telescope, a PA must carefully judge the number of groups that he or she sends to it. Too many groups, and the telescope will be overloaded. Too few, and it will be under-loaded (and will probably thrash). Thus, with improper loading, observations will either not be accomplished at all or will be accomplished at a greater airmass than is strictly necessary.

To date, each of GWH's automatic telescopes has been dedicated to a specific, long-term observing program. The operating schedule for each telescope has been extensively tuned by GWH to achieve excellent loading and, thus, good overall telescope performance. However, even small changes to the observing program make it very difficult to optimize the loading and scheduling. For multi-user telescopes, and for varied observing campaigns, this sort of extensive manual tuning is impractical.

Another problem with the ATIS operations paradigm is that there is no provision for multi-night loading of observations onto the telescope. Any given ATIS telescope has a set of groups, and this set is active on any given night during the year. The definition of an ATIS group does provide a range of dates on which it can be executed, but within this range, the group will be executed without regard to what has happened in the past (or, indeed, with regard to what might happen in the future). In essence, there is no "memory" between successive observing nights. For example, there is no way in ATIS to express the requirement that a particular group should receive a higher priority on some given night if that group has not been observed as desired on a previous night. To accomplish this, the PA would have to manually tune the priorities of all groups from night to night. Thus, it is clearly impractical to provide any long-term loading capability for astronomers if that capability has to be manually implemented by the PA.

4. Problems solved by the APA

It is possible to improve upon the default ATIS group selection rules by using more sophisticated scheduling techniques (Drummond, Bresina, and Swanson, 1994). Specifically, it is possible to improve the quality of observing data by more precisely scheduling groups so that observations are taken at lower airmass (on average), and so that observations are obtained at astrophysically interesting times. Additionally, for a multi-user telescope, better scheduling can result in a fairer allocation of telescope time to requesting astronomers. By implementing a database of observing requests and some historical information regarding their execution history, it is also possible to more effectively load the observations over multiple nights according to astronomer-defined constraints and preferences.

We are building a system to achieve these goals. Our system is called the Associate Principal Astronomer, or APA. It is an "associate" to the PA in the sense that it is responsible for handling the low level day-to-day details of ATIS group management. It is responsible for collecting observing requests (expressed in ATIS), for loading those requests over multiple nights, for scheduling the requests within each night, for monitoring their execution throughout each night (and for rescheduling when things go wrong), and for returning the resulting data to astronomers as soon as it is available. The goal of our project is to have the APA provide automation support for all aspects of the management of ATIS-compatible telescopes. Our focus is on providing software tools to help a PA who represents a community of participating astronomers.

The APA is based on ATIS93 (Boyd, et al., 1993), which represents a significant extension over ATIS89 in a number of ways. We were invited to be part of the International Astronomical Union ATIS93 standardization committee to assist with ATIS extensions in support of advanced scheduling. In collaboration with other committee members, we designed a new group selection advice statement (see Bresina, et al., 1994). This new statement is used to override the default ATIS group selection rules. The committee also agreed on a mechanism for communication with an ATIS-based telescope controller in terms of partial input and partial output files. Together, these new features make it possible to implement a non-native (i.e., external) scheduler that can effectively drive an ATIS93 telescope controller to better serve the scientific objectives of participating astronomers. Using the new capabilities of ATIS93, the APA is able to effectively schedule for and communicate with remotely located automatic telescopes.

5. A day and a night in the life of the APA

In essence, the APA is a three-way moderator, moving information between the telescope, the PA, and the telescope's users. The right half of figure 1 and the following scenario illustrates "a day and night in the life of the APA" to help explain the nature of the APA's role.

5.1. From an astronomer's perspective

There are basically two ways that an astronomer can submit an observing request to the APA. First, an astronomer can use the world-wide web ("the web") to create and submit a request.[1] The APA provides a web "form" that an astronomer fills out; this form allows the astronomer to define his or her groups and to then submit them to the APA's database. We are using a version of George McCook's *create* program (McCook, 1991) as a back-end for these forms. Our system captures the user's form-based specification and converts it into input that is appropriate for *create*, which in turn converts the specification into pure ATIS instructions.

The form-based interface allows each astronomer to list the requests that he or she has in the APA's database, to access the data generated by the execution of any given group, and to delete (or edit) any given group that he or she owns. Alternately, an astronomer can use basic Internet e-mail to send a message to the APA. Such messages can contain requests to add a group to the database, to list the groups that are contained in it, to delete a group contained in it (which must be owned by that astronomer, of course), or to access the data produced by the execution of any given group. For simplicity, our discussion assumes the electronic mail method of group submission, since it is the more common of the two.[2]

Suppose that an astronomer creates an ATIS93 observation request file and sends it via e-mail to the APA. The mailed file is automatically received and parsed to check for syntax errors. If the file adheres to the ATIS93 specification, then the APA e-mails a message back to the astronomer acknowledging successful receipt of the request; otherwise, a message is e-mailed back identifying the request's syntactic errors. Once the request has been accepted, the astronomer is free to go about his or her day-to-day business, while the APA and the telescope take care of the newly submitted requests. At the end of each observing night, the APA e-mails the astronomer a message containing pointers[3] to the data files

---

[1] The web is to documents as the Internet is to computers. While the Internet provides a mechanism for connecting a number of computers together through a common protocol, the web provides a means of linking together hypertext documents that reside on different computers located throughout the Internet.

[2] Access to the web requires a direct connection to the Internet via TCP/IP; e-mail access requires only a gateway of some sort.

[3] Each file pointer is actually a Universal Resource Locator, or URL. A URL is a compact way of naming a protocol, a computer, and a file that resides in the directory system of the given computer. FTP is one possible protocol, but the protocol used by the web is known as HTTP, which stands for HyperText Transfer Protocol.

that were generated by the execution of his or her requests, along with the results necessary for data reduction and data quality assessment.

5.2. From the PA's perspective

The PA is responsible for scientific aspects of telescope use, while the APA is responsible for telescope administration. The PA's real expertise is in ensuring that the telescope is performing useful scientific observations, and in setting the general policy by which the APA schedules observations. The APA provides a number of displays and analyses to help the PA determine whether or not the telescope is performing useful science. The APA does not, however, attempt to offer any advice in this area, since such judgments are based on intuition and many years of scientific experience. In addition to scheduling, the APA takes care of ATIS bookkeeping, syntax checking, accounting, and distribution of observing results.

The APA divides the overall problem of group scheduling into two subproblems: first, it assigns a group to execute on a specific set of dates; second, for any group that has been selected for execution on the current date, it assigns specific times throughout the night at which that group should be executed. The first process is called *telescope loading*, and the scope of time that it covers includes many months. The second process is called *night scheduling*, and the scope of time that it covers includes the hours, minutes, and seconds that occupy a given night. After performing group loading and night scheduling, a new combined ATIS93 file is automatically assembled by the APA and sent to the telescope.

The PA can check how the telescope controller will execute the new request file by displaying a prediction of telescope behavior for the night. This prediction is based on the best schedule found by the night scheduler. The APA's predictions regarding how this schedule will be executed assume that nothing will go wrong during the night. Of course, errors in execution do occur: a cloud obscuring a star can make a group abort, for instance, and an abort causes an execution failure for the current schedule. (Dynamic rescheduling deals with this problem, and is discussed below.) If the PA is satisfied with the nominal schedule, then nothing more needs to be done. If the PA is not satisfied with the nominal schedule, then the APA's scheduling policy can be modified and the scheduler re-run. (Various interface controls are available to the PA that define the APA's scheduling policy.)

The next morning, the results from that night's observations are already stored at the APA, and astronomers who have obtained new data have already been notified via e-mail. If the PA wants to assess the quality of the night's observation schedule and results, the actual telescope behavior can be displayed. Once the PA has tuned the APA to consistently produce high quality schedules, the APA takes care of routine observation loading and scheduling with only occasional supervision. The PA must remain vigilant with respect to scientific quality, however. The APA makes no attempt to ensure that the observing results are serving the needs of the telescope's users. Such scientific and engineering judgment is left firmly in the hands of the PA.

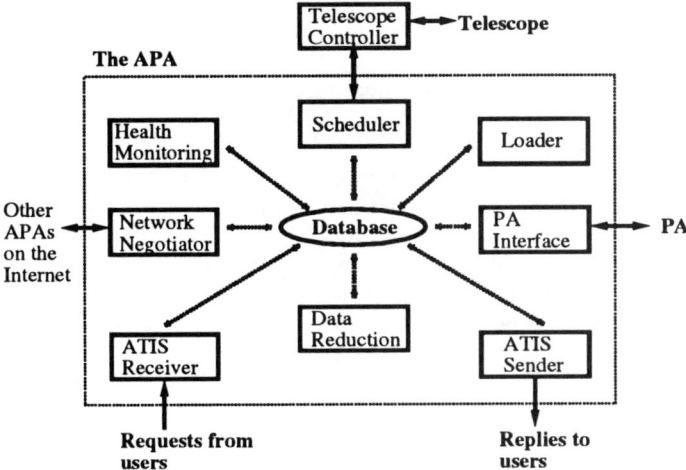

Figure 2. Block diagram architecture of the APA.

5.3. From the telescope controller's perspective

Just before nightfall, the ATIS93 input file is automatically transferred to the telescope controller along with the observation schedule. The controller executes the schedule and, at the end of the night, transfers the ATIS93 output file back to the APA. Such an interaction is ideal but rarely realized, since the schedule can break during execution. Schedule breakage is a result of unpredicted (sometimes, unpredictable) events occuring during execution, such as a cloud obscuring a star during automatic acquisition. The ATIS93 specification was designed with this in mind, however. ATIS93 allows for partial input and partial output files to be transmitted between the APA and telescope controller during the night. The partial output files enable the telescope behavior and status to be monitored during the night – either by a person (for example, to check the status of the telescope mechanics and optics) or automatically by the APA. Partial input files enable the APA to transmit new schedules and new groups during the night as necessary.

6. The architecture of the APA

This section provides an overview of the architecture of the APA. It covers the basic mechanisms by which the APA connects together the community of astronomers, the PA, and the telescope.

6.1. Architecture components

Figure 2 shows a block diagram for the architecture of the APA. Central to the APA is the database of ATIS requests, shown in the middle of the figure. This database contains all users' ATIS requests, together with the data that these requests have generated. It also contains information regarding the telescope itself: filter set, observing window, latitude, longitude, elevation, PA scheduling preferences, and other relevant parameters to allow for effective telescope management.

The figure also shows a number of components that extract data from, and enter data into, the database. Included among these components are the scheduler, loader, PA interface, ATIS sender, data reduction, ATIS receiver, network negotiator, and health monitoring.

Briefly, the eventual goal for the network negotiator is to have the APA be able to work with other copies of itself on the Internet, handing off and accepting observing requests as conditions warrant.[4] In terms of health monitoring, the goal is to have the APA able to spot long-term trends in telescope behavior and to make recommendations regarding possible mechanical, optical, and electrical faults, together with suggestions for their correction. It should be possible to integrate existing work on automated telescope monitoring (for instance, see Monahan, Patterson-Hine, and Iverson, "Automated Telescope Monitoring and Health Diagnosis", this volume). Such integration is an interesting area for future work.

For now, the network negotiator and health monitor functions are simply place holders and are discussed no further. We also do not say much about the data reduction function shown at the bottom of the figure; for now, we are using algorithms written by one of us (GWH) that are specific to a particular telescope and observing program. These algorithms have been ported by Don Epand (of Fairborn Observatory) to run under Unix, and we have simply incorporated his work into the APA. Don Epand is working on a more general version of GWH's data reduction system, and we hope to incorporate it into the APA as and when appropriate. Thus, the focus of this discussion is on the ATIS receiver, scheduler, loader, PA interface, ATIS sender, and on the database that binds them together.

6.2. Activation, loading and scheduling

The basic process within the APA is a clocked loop that counts down to sunset at the telescope site. We refer to this clocked loop as the *activator*, since it is responsible for activating various parts of the APA at appropriate times of the day and night. At a fixed time before sunset at the telescope, the activator "freezes" the database. During the day, requests are received from the telescope's community of users by the ATIS receiver, parsed to ensure validity, acknowledged, and then incorporated into the database. Freezing the database means that any request received after this time will not be considered for execution during the coming night. When the database is frozen, users are still able to submit requests, of course; requests received after the database is frozen for the current night are simply not considered for execution until after the database is unfrozen just after dawn.

Once the database is frozen, the activator starts the loader in order to determine the nights on which each group should be considered for execution. The loading process is sketched graphically in figure 3. The loader examines the telescope's execution history (contained in the database) to see when each group ran last. It also examines execution constraints that have been imposed

---

[4] Using multiple copies the APA distributed over a network, it should be possible to implement a globally coherent observing system of the sort outlined by Crawford (1992).

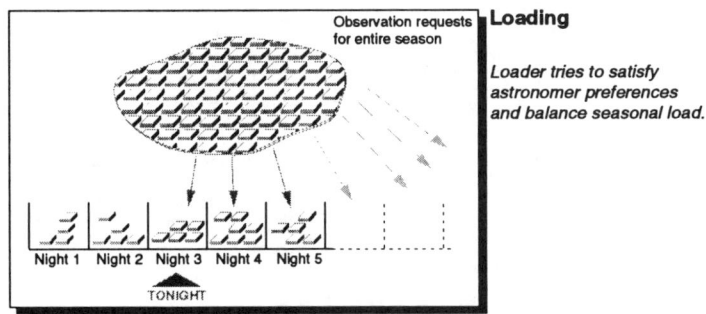

Figure 3.   Assignments made by the loader.

by each astronomer on his or her ATIS groups. These constraints take the form of metric temporal expressions on the execution count and inter-execution gap length. For instance, an astronomer might specify that she is interested in no fewer than 10 observations, that she requires no more than 20, and that each observation in the sequence should be scheduled every other night. The loader takes into account all such execution constraints, posed by all astronomers, and attempts to find an assignment of groups to nights that satisfies all constraints and astronomer preferences as much as possible. The details of this process are beyond the scope of this discussion; see Bresina (1994) for more information.

The result of the loading process is an assignment of groups to nights, stretching for months into the future. The coming night is, of course, the first night in this sequence. The activator process next starts the night scheduler in order to find a good sequence in which to execute the coming night's groups. The scheduler searches over various possible execution sequences, eventually selecting one that scores well according to various PA-defined attributes. This process is shown graphically in figure 4. The scheduler is given a mathematical statement of what constitutes a good schedule. This mathematical statement is called an *objective function*, and is a weighted sum of various schedule attributes. The scheduler uses the objective function to search a space of possible schedules. The objective function is used as a heuristic to guide the search – it allows the scheduler to make informed tradeoffs about which particular group to execute at which particular time.

It is important to note that the interaction between the loader and the scheduler involves some feedback, since the loader evaluates the potential performance of the scheduler on any given potential load. This feedback allows the loader to form a set of groups for each night that the scheduler can effectively sequence. However, the details of this process are beyond the scope of this discussion; see Bresina (1994).

The result of scheduling is a set of ATIS93 instructions that tell the telescope controller the order in which to execute the given set of groups (these ATIS93 instructions encode the schedule). The APA assembles a file that contains the selected groups for the coming night, together with a schedule that dictates how they should be executed. More detail on these ATIS93 instructions and how they are generated by the scheduler is available elsewhere (Bresina, *et al.*, 1994).

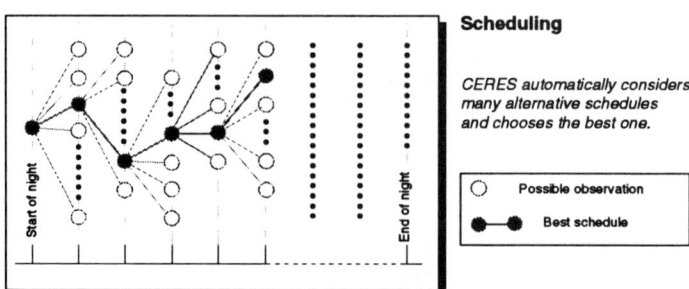

Figure 4. The scheduler's search space.

6.3. Schedule execution and dynamic rescheduling

At a pre-specified time, the telescope controller takes the schedule file from the APA via FTP.[5] The semantics of the schedule is defined in the ATIS93 standard. Thus, any telescope controller that understands ATIS93 is able to execute an APA-produced schedule. At the appropriate time, the telescope controller rolls off the observatory roof (or opens the dome) and starts executing groups as specified by the schedule.

The telescope controller is able to send partial ATIS93 output files to the APA throughout the night. The presence of a partial output file is detected by the APA, and, in general, is taken to mean one of two things. Either, the partial output file just represents a report on the telescope controller's current status; in this case, the APA simply updates any active tracking displays and continues watching for files, since no particular response is required by the controller. Alternatively, the partial output file can represent a schedule break, in which case the telescope controller is expecting the APA to produce a new schedule. The APA can easily distinguish between the two cases by examining the contents of the partial output file. If the presence of the partial output file represents a schedule break, then the waiting scheduler is given the break information and is asked for a new schedule that covers the remainder of the night (the scheduler takes into account what has already been executed up to the break). The APA and the telescope controller continue through the night in this manner, passing ATIS93 partial input and output files back and forth, communicating current status and new schedules as required.

While this dynamic rescheduling can handle execution errors, the APA goes one step further, attempting to avoid such errors when possible. The APA uses a technique called Just-In-Case scheduling to form robust schedules. A robust schedule is one that is less likely to break during execution. The Just-In-Case technique forms contingent schedules to cover likely schedule breaks; for more information on Just-In-Case scheduling, see Drummond, Bresina, and Swanson (1994).

If, for some reason, the communication link expires, then the telescope controller "falls back" on its native ATIS group selection rules. This scheme allows the telescope controller to continue executing the given set of ATIS groups, even

---

[5] This transfer can be accomplished via other protocols and communications mechanisms such as Kermit, if necessary or desired.

without the benefit of the APA's scheduling advice. We call this the "principle of independent competence", since the telescope controller is able to schedule and execute groups with or without the help of the APA. Of course, schedules produced by the APA are typically much better than those produced by the ATIS group selection rules. However, we feel that it is better for the controller to go ahead and execute some groups, rather than sitting idle, waiting for a better solution that might never arrive.

In the morning, the APA examines the database to see which astronomers have obtained new data. An e-mail message is composed for each such astronomer, informing him or her of how to obtain the data. The message is composed and sent by the ATIS sender component of the APA architecture (shown in the lower right hand part of figure 2). The activator process then queues itself to run again the next evening, at which time the entire procedure starts again.

6.4. Displays and controls for the PA

The APA provides a number of interface tools so that the PA can examine and customize the process described above. These interface tools provide a means for the PA to track down execution problems, to examine the contents of the ATIS database, to watch the telescope's behavior in real-time, and to tune the scheduler's search engine.

The execution analysis tool provides a set of routines for examining ATIS output files in order to find patterns of successful executions and aborts. These tools extract data from ATIS output files over a range of dates and plot that data in a manner specified by the PA. For instance, the PA can request a plot of abort types in terms of star magnitude versus star color index; such a plot shows the correlations that exist between abort type (the reason for a failed execution) and the characteristics of the star being studied. (See the APA project web page for some example displays.)

The APA also provides graphical displays of the contents of the ATIS database, allowing the PA to see the users' requests in a variety of different formats. There are also a couple of tracking displays that are updated automatically during schedule execution through the night. One school of thought suggests that these tracking displays represent a tremendous retrograde step, since one of the goals of ATIS-based operations is to let all astronomers (including the PA) sleep at night while the telescope is busy collecting data. However, we feel that such tracking displays should be useful for detecting problems in real-time, and hopefully, will be used only every now and then.

Finally, we have plans for a class of interface tools that allow the PA to tune the APA's scheduling search engine. These tools are not yet implemented, but our plan is to make it possible for the PA to set the weights on all terms within the scheduler's objective function. Adjusting the weights of terms within the objective function is a way of expressing relative preference over the different attributes of a good observing schedule. It is still too early to say how successful this sort of tuning mechanism will be, but we eagerly look forward to completing our first implementation and subsequent testing.

6.5. Tools used to construct the APA

Our approach in constructing the APA has been to use high performance, free, off-the-net software. We have done this to ensure that the APA can be freely distributed and modified by others. The ATIS parser has been implemented using variants of Lex and Yacc, Unix tools for building lexical analyzers and compilers. As a result, it should be easy to modify the parser to work with a new version of the ATIS language (whenever a new version is defined), or to work with another telescope control language altogether. The ATIS database has been built on top of the Gnu database management (GDBM) system.[6] The interface is written entirely in Tk, a library of graphical tools that works with the X11 windowing system. The framework that holds all the APA utilities together is written in Tcl, the tool command language, designed to work well with Tk.[7] (See Ousterhout (1994) for more information on Tk and Tcl.) The core of the scheduler is written in "C", with surrounding support routines written in Perl (Wall and Schwartz, 1994) and Tcl.

7. Implementation and testing status

In the spring of 1994, we first implemented and tested a LISP-based version of the APA's scheduler. Testing of this system was carried out on a telescope simulator, and on a 10" telescope located in Ft. Collins, Colorado. These early tests demonstrated that it is possible to improve upon the scheduling offered by the default ATIS group selection rules. Various astronomers (including one of us, GWH), also made it quite clear that scheduling was only part of the overall story, and that if our goal was to produce a useful telescope management tool, we would have to address the entire life cycle of an ATIS request, from first submission to final release of the data. We also realized that LISP, powerful as it is for prototyping, is not the language of choice among astronomers. Thus, we made the commitment to redesign the entire system in the more commonly accepted "C" language.

The complete system is still undergoing construction, but most parts of it now exist. We have implemented the ATIS parser and database and have built a graphical tool that allows a user to employ the parser to construct a database from raw ATIS files. We have built a suite of execution analysis tools and are currently evaluating them on live data obtained from the automatic photoelectric telescopes located on Mt. Hopkins (the telescopes are managed by Louis Boyd of Fairborn Observatory). The new C-based version of the scheduler is almost complete. Preliminary tests suggest that it is able to generate a schedule for an entire night in about 60 seconds. This performance is perfectly acceptable for deployment on a real telescope, but we must exercise the scheduler in a closed-loop dynamic rescheduling context before making any specific performance claims. To this end, we are collaborating with Fairborn Observatory on

---

[6] GDBM is available from the Free Software Foundation. A place to look on the web is http://csugrad.cs.vt.edu/manuals/gdbm/gdbm_toc.html .

[7] Tk and Tcl are available from various FTP sites around the Internet, primarily from the University of California at Berkeley. On the Internet, see ftp://ftp.cs.berkeley.edu/ucb/tcl/ .

a new ATIS93-based telescope. This telescope will be located at Fairborn, and its controller will implement enough of the ATIS93 command set to allow it to execute APA-produced schedules and to allow the APA and telescope controller to communicate throughout the night (as discussed in the previous section). The loader currently exists in prototype form only, written in LISP.

We have also developed a set of prototype web forms that allow an astronomer to specify observing campaigns. These forms are based on a proposal by Louis Boyd (Boyd, *et al.*, 1993) for a goal-directed observation specification language. These forms are also connected to a version of McCook's *create* program (McCook, 1991), in order to allow an astronomer to specify the groups that are part of any given campaign.

8. Discussion

The overall goal of our project is to provide automation support for the management and use of remotely located, automatic telescopes. So far, we have focused on building the core of an Associate Principal Astronomer, or APA. This core consists of an ATIS parser and database, automatic group loading and scheduling mechanisms, together with a means for executing schedules automatically and dynamically rescheduling as necessary. Experience gained with simulation tests and preliminary runs on a single automatic telescope have been very encouraging.

It is clear that there are other instances of the fully automatic execution paradigm. Others have built APA-like systems (Richmond, Treffers, and Filippenko, 1993). The primary advantage of the APA is that it uses advanced loading and scheduling techniques and that it operates with any telescope that adheres to the ATIS93 standard.

Of course, NASA has a number of orbital telescopes that are operated remotely. The Hubble Space Telescope (HST), for instance, is operated in a way that is somewhat similar to our APA paradigm. However, there is a significant amount of human infrastructure associated with the management of HST. Such infrastructure is expensive, and it cannot be replicated for every single telescope that is to be run automatically. Clearly, the human infrastructure surrounding HST performs useful tasks that our APA paradigm ignores: for instance, helping users formulate their telescope requests and helping users make sense of the data they obtain. Our APA paradigm leaves all such tasks firmly in the hands of the telescope users (and their scientific community).

Additionally, since HST is an orbiting telescope, and since it costs many millions of dollars, the human infrastructure is well-justified: nothing can be allowed to go wrong with the telescope, or a lot of money and time will be wasted. The telescopes for which the APA paradigm has been developed are rather different: they tend to be cheap (on the order of hundreds of thousands of dollars, not millions), and they are bolted to a platform with essentially infinite inertia (the Earth). Thus, many of the delicate operations issues that arise for HST simply do not come up for our telescopes. A human infrastructure the size of that required for HST is clearly unreasonable for one of our target telescopes. Our APA operations paradigm requires one workstation (or a high-end personal computer), one experienced astronomer to act as the telescope PA, and an engineer to fix the telescope and observatory control systems when things go wrong.

The goal of this paper has been to provide an overview of the APA operations paradigm, and to explain, in general terms, how it is being built. We hope to soon report on specific results obtained by using the APA to manage a number of remotely located, fully automatic telescopes.

Acknowledgments

Thanks to Bill Borucki, Louis Boyd, Othar Hansson, Butler Hine, George McCook, Ann Patterson-Hine, and Boris Rabin. Thanks also to Mel Montemerlo (NASA Headquarters) and Peter Friedland (NASA Ames) for making this entire project possible. GWH acknowledges support from NASA Ames grant NCC 2-883.

References

Bresina, J., Drummond, M., Swanson, K., and Edgington, W. 1994. Automated Management and Scheduling of Remote Automatic Telescopes. In *Optical Astronomy from the Earth and Moon*. D.M. Pyper and R.J. Angione (eds). ASP Conference Series, Vol. 55.

Bresina, J. 1994. Telescope loading: a problem reduction approach. In *Proceedings of the Third International Symposium on Artificial Intelligence, Robotics, and Automation for Space*. pp. 351–354. Jet Propulsion Labs. Pasadena, CA. (JPL Publication 94-23)

Boyd, L., Epand, D., Bresina, J., Drummond, M., Swanson, K., Crawford,D. Genet, D., Genet, R., Henry, G., McCook, G., Neely, W., Schmidtke, P., Smith, D., and Trublood, M. 1993. Automatic Telescope Instruction Set 1993. *International Amateur Professional Photoelectric Photometry IAPPP Communications*, No. 52, p. 23.

Crawford, D. 1992. GNAT: Global Network of Automated Telescopes. In *Automated Telescopes for Photometry and Imaging*, S. Adelman, R. Dukes, and C. Adelman (eds), ASP Conference Series, Vol 28, p. 123.

Drummond, M., Bresina, J., and Swanson, K. 1994. Just-In-Case Scheduling. *Proceedings of the 12th National Conference on Artificial Intelligence*. Seattle, WA, AAAI Press.

Drummond, M., Bresina, J., and Swanson, K. 1994. Robust Scheduling and Execution for Automatic Telescopes. In *Intelligent Scheduling*, M. Zweben and M. Fox (eds). Morgan Kaufmann, San Francisco, CA. pp. 341–369.

Emerson, D.T., and Clowes, R.G. (eds.) 1993. *Observing at a Distance* (Proceedings of a Workshop on Remote Observing). World Scientific Publishing, River Edge, NJ.

Genet, R.M., and Hayes, D.S. 1989. *Robotic Observatories: A Handbook of Remote-Access Personal-Computer Astronomy*. AutoScope Corporation, Ft. Collins, CO.

McCook, G.P. 1991. Create: An ATIS Management Program. In *Advances in Robotic Telescopes*, M. Seeds and J. Richard (eds). Fairborn Press, Mesa, AZ. p. 263.

Ousterhout, J.K. 1994. *Tcl and the Tk Toolkit.* Addison-Wesley Publishing Company, Menlo Park, CA.

Richmond, M., Treffers, R., and Filippenko, A. 1993. The Berkeley Automatic Imaging Telescope. *Publications of the Astronomical Society of the Pacific.* No. 105, p. 1164.

Wall, L., and Schwartz, R. 1994. *Programming Perl.* O'Reilly and Associates, Inc. Sebastopol, CA.

AUTOMATED TELESCOPE MONITORING AND DIAGNOSIS

CHRISTINE M. MONAHAN
Recom Technologies
NASA Ames Research Center
M S 269-4, Moffett Field, CA 94035-1000

F. A. PATTERSON-HINE   DAVID L. IVERSON
NASA Ames Research Center
M S 269-4, Moffett Field, CA 94035-1000

ABSTRACT   This paper discusses automated telescope monitoring and diagnosis work currently being performed at NASA Ames Research Center in cooperation with AutoScope Corporation. Topics addressed include information acquisition from the telescope control system, techniques for system monitoring and diagnosis of failures, and the software tools used in the implementation of a monitoring and diagnosis program for a telescope subsystem.

1. INTRODUCTION

Use of untended, fully automatic telescopes bring the need for automated health management for the telescope control system. Automated monitoring of the telescope control system(TCS) allows astronomers to receive data about the TCS without the performance of manual tests by an onsite engineer. Automated fault diagnosis provides the cause of a failure should one occur. Research in this area is currently being performed at NASA Ames Research Center. The primary goal of this research is to provide automated monitoring and fault diagnosis capabilities for remotely operated telescopes to reduce costs and down time. The cooperative effort between Ames and AutoScope Corporation will allow technologies and system models developed at Ames to be transferred to industry. This approach is useful at inhospitable environments such as remote mountain tops or the South Pole, and will be imperative at the planned Lunar Outpost where the luxury of on site engineers is not only impractical but impossible. This approach will allow telescopes to be repaired by technicians, in the case of the Lunar Outpost, astronauts.

2. OVERVIEW

Autoscope's TCS-200 Telescope Control System (TCS) was modeled using digraphs. Digraphs are a type of reliability model that displays the way failures propagate through a system. Closely resembling the schematic, the digraphs of the TCS emulate the functional flow through the system. For example, the failure of the 24 volt power supply is known to cause the failure of the stepper driver function. Therefore, its digraph representation will have the "24 volt power supply" node connected to the "stepper driver" node to indicate this dependency, as depicted by the digraph in figure 1.

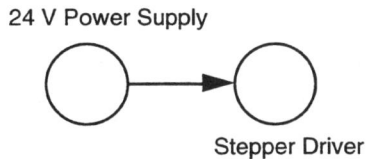

Fig. 1. Simple Digraph Model.

To utilize these digraphs for failure diagnostics, system status will be obtained from AutoScope's Automatic Diagnostic System, the ADS -100. The ADS-100 provides data on 37 key test points within the telescope control system. This enables us to define points in the digraph where component failures could be detected. These points in the digraph are referred to as observation nodes because they represent places in the system being modeled that the user may observe a failure according to test results. The observation nodes will be set as failed according to the findings of the ADS-100. For example, one test point is at the 5 volt power supply. If the ADS-100 finds by sampling the test point that the voltage is not within acceptable limits, the observation node for that test point will be set as failed. This allows the digraphs to be used for near real-time failure diagnostics by propagating the effects of hypothesized failures along the digraph to find component failures which could cause the observed conditions.

In order to utilize these constructed digraphs for automated fault diagnosis, several software and hardware packages must be integrated. Figure 2 displays the data flow through the major components.

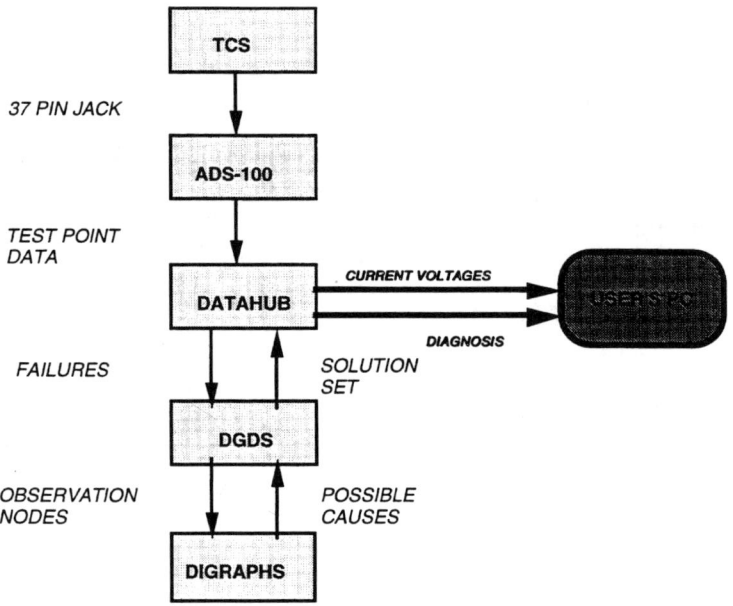

Fig. 2. Data flow through major components.

The following sections discuss the major components involved in the automated monitoring and diagnosis application, with specific emphasis on the digraph models.

3. DIGRAPH MODELS

Digraph models display the way failures propagate through a system. The models consist of nodes and AND gates connected by directed edges. Each digraph node represents a failure in the modeled system. The digraph edges show how the occurrence of a failure can flow through the system to cause other failures. If a node is marked as failed, the failure will propagate through all directed edges leading out of that node and mark any connected nodes as failed. Redundancy is modeled with digraph AND gates. Both input nodes of an AND gate must be marked as failed in order for the failure to propagate past the gate and affect the node on the other side. In a graphical digraph depiction, AND gates are drawn as bars and regular nodes are drawn as circles (see fig. 1). Digraph nodes can be in one of two states, true or false. If a node is true, or *marked*, it means the failure that the node represents has occurred. If the node is false, or *not marked*, then the failure has not occurred.

Digraph models can be derived from system schematics in a fairly straightforward manner by associating a digraph node with each component in the schematic, adding directed edges to represent physical connections, and augmenting that basic digraph with knowledge about component failure modes and system design.

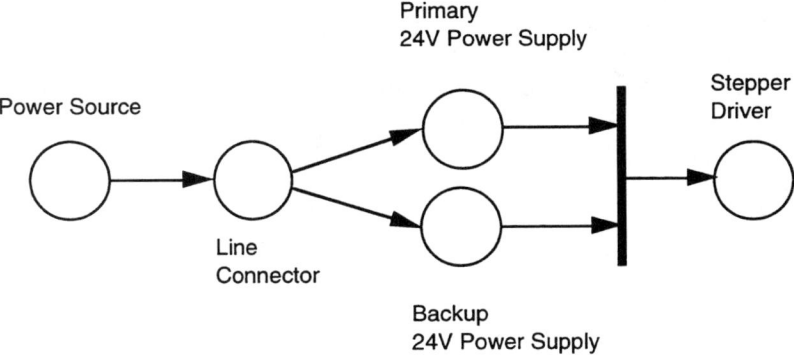

Fig. 3. Stepper Driver Digraph.

Figure 3 shows a simple, hypothetical digraph model of part of a telescope control system. Each node represents a failure of a component of the system and the edges correspond to the flow of power through the system. The stepper driver supplies current to the windings in the stepper motor (linear actuator). Notice that a failure in the power source or the line connector could propagate and cause the power supplies to fail in their function of delivering power to the stepper driver. The bar in the digraph is an AND gate which indicates that both the primary and backup power supplies for the stepper driver must fail to operate before the stepper driver fails due to lack of power. Depending on the desired level of detail, additional nodes (e.g., fuses, switches, connector pins) could be added to this digraph.

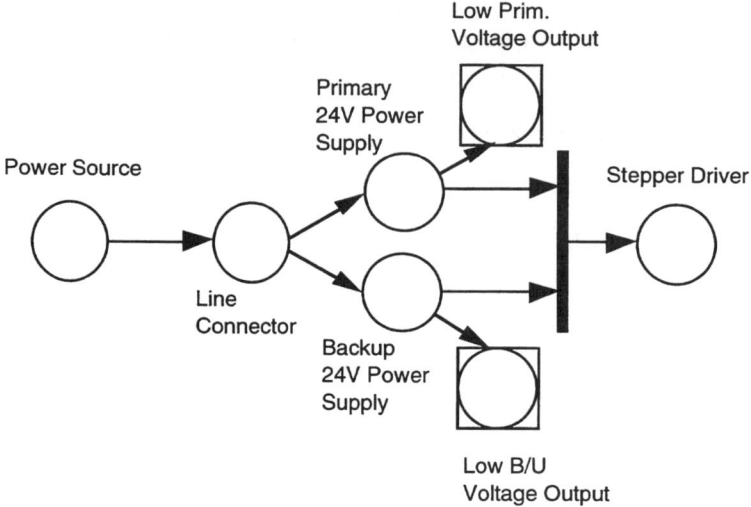

Fig. 4. Stepper Driver Digraph with Observation Nodes (square outline).

In order to use digraph models for fault diagnosis we must include points in the digraph where information regarding the health of the TCS is available. Points where actual system information is available is modeled using observation nodes. Observation nodes correspond to out-of-limits sensor conditions. These observation nodes are added to the digraph as outputs of failures that could cause the sensed condition. In terms of the TCS, these are test points where voltage levels are measured and found to be out of acceptable limits. Figure 4 shows observation nodes that might be added to the stepper driver digraph shown in figure 3. In this example, two observation nodes have been added and are indicated by the nodes with the square outlines. These new nodes correspond to test points with in the TCS where voltage levels for the power supplies can be sampled.

The following sections describe the application that will use the digraph model technique described above for fault diagnosis.

4. TCS DIGRAPH MODELS

The actual digraph models of AutoScope's TCS-200 are much larger and more complex that the previous two digraph examples. Much like the system schematic, the digraphs proceed from the Telescope Control

Computer (TCC) out to the major components. The nodes represent functional failures that can be experienced along the path. For example, as a signal is sent to the linear actuator, there are a number of problems which could be encountered along the way. The stepper driver, its power supply, connector pins, or cables located between the stepper driver and the linear actuator could fail. Hardware failure of one of these components would result in the failure of the linear actuator to perform its task. In the digraphs, this is represented with a node for each of the components. The components are linked together in a string from the TCC through the stepper driver and power supply out to the linear actuator. Thus the functional flow and dependencies are included in the model.

Information about AutoScope's TCS-200 telescope control system was obtained from the system schematic, electrical diagrams, technical manual, and AutoScope's principle engineer. Once sufficient information about the design of the TCS-200 was obtained, the TCS digraph models were constructed using the Digraph Editor, a software package developed for NASA by Lockheed Engineering and Sciences Company. Each piece of the digraph model was checked by the principle engineer to ensure the correct functional flow and dependencies were captured. The final digraph model of the TCS-200 telescope control system consists of hundreds of nodes. Each node is represented by a circle with a text block identifying it. The text block consists of a 16 character mnemonic, a component description, and a schematic link. The 16 character mnemonic is used by the digraph processor in its analysis. The first character in the mnemonic represents the node type. "H" stands for hardware node, "F" for function node, and "O" for observation node. The next twelve characters represent the component. For example, "TCS_J106_P25" stands for jack 106 pin 25. The last three characters in the mnemonic represent the failure mode. "_00" stands for "fails out of limit," "_01" stands for "fails," and "_90" stands for "fails to function." The component description part of the text block simply makes the digraphs easier to read; it plays no functional role in solving the digraphs. The schematic link provides a link between a node and its corresponding component in the system's schematic.

The digraph in figure 5 represents the possible TCS failures that affect the instrument selector function; it is a small portion of the actual TCS-200 digraph. The failure of the instrument selector to perform its task is represented by the node on the far right. The nodes representing the failure of the 5V and 24V power supplies are represented in by the nodes with triangular outlines. This outline is simply an indication to the modeler that this node is repeated elsewhere within the digraphs. As seen in the digraph, a signal is sent from the 24V power supply to the instrument selector and out to the output modules for the four ports of the selector. The level of detail with which the TCS-200 digraphs were constructed can also been seen in this example. Between the 24V power supply and the instrument selector, the model includes hardware failures on the connector pin level, this level of detail is maintained between the instrument selector and each of the four output modules. Each output module is additionally dependent on the 5V power supply. Therefore, as indicated by the digraph, any of the four output modules would fail to perform its task if either the 5V or the 24V power supply failed.

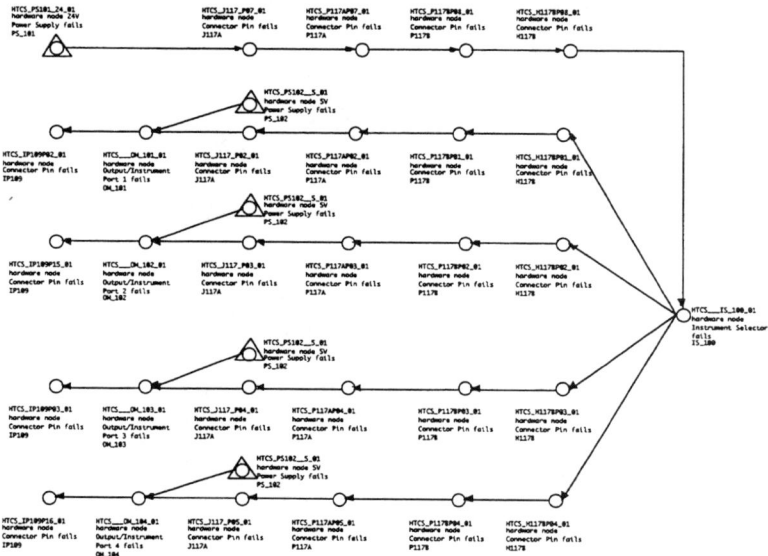

Fig. 5. Digraph Model of the TCS Instrument Selector Function.

5. DATAHUB

Once the ADS-100 has sampled the test points within the TCS, DataHub, a software package developed at Ames, can determine and display which limits have been maintained and which have been exceeded. DataHub is a UNIX based C program running with X-Windows. The TCSView portion of DataHub provides the user with the telescope control system schematic. Red buttons at key points in the schematic depict the test points. Clicking on one of these buttons creates a new window which displays the acceptable voltage range and the current voltage level for that particular test point as obtained from the ADS-100. DataHub also allows the status of all the test points to be viewed simultaneously. This allows the astronomer to continuously monitor the

TCS without an on site engineer manually performing tests. Viewing the voltage levels straight from the TCS without operation interruption saves time and money. Furthermore, the display of the test point data over time allows trends to be detected and possibly a fix before a failure occurs.

6. DIGRAPH DIAGNOSIS SYSTEM

Those test points which have failed out of limit will be marked as failed in the digraphs via the Digraph Diagnosis System (DGDS). DGDS is a UNIX based C program also developed at Ames. This digraph solution algorithm marks the failed test points in terms of observation nodes within the digraphs. It then solves the digraphs by backward propagation of the failure to determine the possible causes for the observed set of failures. To see how this is useful, look back at stepper driver digraph in figure 4. Suppose we know from the system data that the primary 24V power supply has failed to be within acceptable limits. By backward propagation it can be determined that this failure could be due to the failure of the 24V power supply, the line connector, or the power source. However, if the test point for the backup 24V power supply yields an acceptable voltage then we know the observation node is considered good. Therefore, also by backward propagation it can be determined that the back up 24V power supply is functioning correctly as well as the line connector and power source. Thus, the fault has been isolated to the primary 24V power supply.

After fault detection and isolation, the nodes which may have caused the observed failure symptoms will be displayed to the astronomer. This failure diagnosis will be viewed by the astronomer through the DataHub package.

7. CONCLUSION

The digraph model is a graphical representation of a system using nodes, edges, and-gates, and or-gates. The failure of system components are represented by nodes, which are linked together in logical failure sequence by lines and arrows(edges). Digraph models of Autoscope's TCS-200 will be used as part of a automated monitoring and diagnosis system. Automated information acquisition will provide researchers with the status of the control system health without performing manual tests. Furthermore, the integration of several software programs together with the TCS digraphs will enable automatic fault diagnosis.

8. FUTURE WORK

Interfaces between the ADS-100, DataHub and DGDS are yet to be developed. Integration of DGDS and DataHub is planned for the end of fiscal year 94. Integration of the health monitoring software and the ADS-100 is planned for the following year, and will be followed shortly there after by testing on site at AutoScope Corporation. Once data from ADS-100 is available via Internet/ATIS93, the monitoring and diagnosis system will be an optional part of the Associate Principle Astronomer (APA) currently being developed at NASA Ames. The future goal of the integration with the APA is to provide a reactive planning system that can reschedule telescope activities to enable the telescope to continue being used in a degraded mode even when component failures occur.

ACKNOWLEDGMENTS

We wish to thank Ken Valentine for his continued support and many hours of helping us to capture some of his engineering knowledge in the form of digraphs.

REFERENCES

Iverson, D. L. and Patterson-Hine, F. A. 1990 in *A Diagnosis System Using Object-Oriented Fault Tree Models*, The Fifth Conference on Artificial Intelligence for Space Applications Proceedings, Huntsville, AL.

Iverson, D. L. and Patterson-Hine, F. A. 1993 in *Digraph Reliability Model Processsing Advances and Applications*, AIAA Computing in Aerospace 10 Proceedings, San Diego, CA.

Patterson-Hine, F. A., Boyd, M. A., and Iverson, D. L.
1992 in *Automated Monitoring and Diagnosis of Telescope Control System*, SALUTE Workshop Proceedings, NASA ARC.

Valentine, K. M. 1994 in *Self Diagnostics of Automated Telescopes,* Optical Astronomy from the Earth and Moon, ed. D. Pyper Smith and R. Angione, ASP Conference Publication Volume 55.

Drummond, M., Bresina, J., Swanson, K., Edington, W., Drasher, E. and Henry, G. 1994 in *The Associate Principle Astronomer,*, The 106th Annual Meeting of the Astronomical Society of the Pacific Proceedings, Flagstaff, AZ.

DIAGNOSTIC ALGORITHMS AND HEALTH MONITORING OF AUTOMATIC TELESCOPES

KENNETH M. VALENTINE
MICHAEL A. DONAHUE
AutoScope Corporation, 2637 Midpoint Drive, Suite D
Ft. Collins, CO 80525
Internet: 73362.2514@compuserve.com

ROBERT W. VALENTINE
University of Arizona, Department of Physics, Tucson, AZ 85719
Internet: bobby@soliton.physics.arizona.edu

ABSTRACT   The paper discusses current diagnostic and health maintenance techniques of AutoScope Corporation's automatic telescopes and accompanying control systems. The first section describes the hardware and software employed by the control system of the automatic telescope. The second section describes specific diagnostic and health maintenance algorithms developed by AutoScope and the NASA Ames Research Center. The final section suggests possible applications for this technology developed by both NASA and AutoScope.

1.0 AUTOMATIC TELESCOPE ENVIRONMENT

The telescope control system referenced to throughout this paper is AutoScope Corporation's Second Generation Control System. Following is a brief description of the overall robotic telescope system.
   Each system rack employs two 386 personal computers connected across an Ethernet Local Area Network (LAN). The Master Computer (MC) directs high level operations of all other computers by issuing command messages across the LAN. The MC handles all external communications for real time control and file transfer via telephone lines or an external control computer (normally a *SUN SparcStation*) connected to the Internet high bandwidth

communications network. The MC controls the CCD camera image acquisition, centering and focus functions.

The second computer is the Telescope and Observatory Control Computer (TOCC) which directs all telescope movement using the Telescope Control Card. This motion control card coordinates all axes of telescope motion, monitors limit and home transducers, and accesses all digital I/O functions for the control system.

Interfaced to these computers are various control system modules. The modular capabilities include dome control, automatic guiding control, observatory control, power control, and telescope control. The telescope control unit is a rack mounted chasssis housing power supplies, digital I/O modules, microstepping motor drives, and several mechanical relays. Each of these module components is fitted with a test lead such that its performance can be evaluated.

2.0 SELF DIAGNOSTICS AND HEALTH MAINTENANCE

Due to the complexity of telescope control subsystems, health maintenance techniques must be developed and implemented by on-site engineers and technicians. Unfortunately, many universities and institutions cannot afford this luxury. Thus, AutoScope Corporation and the NASA Ames Research Center have developed a relatively low cost Automatic Diagnostic System (ADS-100) to retrieve key telescope and control system data. This effort was the first step for self maintenance of a telescope control system. It is important to note that this unit was designed to substantially reduce operational costs and technical support at the telescope site itself.

The Automatic Diagnostic System (ADS-100)

The ADS-100 is a rack mounted analog to digital converter of robotic observatory control system voltage levels. It is the vehicle for acquiring system data from AutoScope's telescope control module, observatory control module, power control module, and automatic guider control module. Each controller's most important signals are monitored as the system is on-line. Any deviations from "normal" operation are brought to the attention of the system's master computer. Of the four previously mentioned controllers, this paper will concentrate on the diagnostic and health maintenance techniques of the TCS telescope control system.

ADS-100 Architecture

The ADS-100 acquires data with the use of an *Analogic DAS 12/50* data acquisition card mounted within the master computer of the control system. It is a high speed analog to digital converter with the ability to sample 256 test points at rates up to 50 KHz. The ADS-100 is a rack mounted chassis containing two multiplexer boards (expandable up to four to maximize channel capacity), a signal conditioning network, and four cable ports for the control system modules.

ADS-100 Software

The ADS-100 software consists of two high level programs residing in the control system's master computer. One program, *CSTEST*, samples each TCS test point as the telescope is orchestrated through a predawn initialization routine. If any major discrepancies detrimental to proper operation of the telescope are encountered, the master computer and principal astronomer will be informed on how to take diagnostic action to rectify any hardware malfunctions. A nightly routine such as this allows continuous operation of a telescope and determines possible failure trends by comparing previously weighted data from earlier initialization routines.

The second software tool, *CSVIEW*, is highly effective for on-site troubleshooting. This program displays all available test points and their instantaneous voltage states. If any voltage parameter exceeds acceptable bounds or nears this condition, the monitor will display these findings using a simple "stoplight" color code. The code displays normal readings in green and out of bounds readings in red. If the signal is within 10% of the boundary conditions, a yellow voltage reading cautions the telescope operator.

ADS-100 Diagnostic Algorithms

The algorithms implemented by *CSTEST* include simple movements in both equatorial axes (three axes of motion for the Alt-Az models) and secondary mirror assembly axes. Home sensing, limit sensing, etc., are software nested within these movement commands such that proper operation of the telescope can be determined. An example of such a movement and its appropriate sensing status is described below.

Telescope movement to each axis' home position is critical for proper operation of the telescope. This is attributed to the dependency of telescope pointing and tracking on a highly repeatable home position. Below is such an algorithm implemented by the nightly *CSTEST* software tool.

```
move_scope to neg_limit
   while not (home_sensed and pos_limit_sensed) then
      begin
         step_scope in positive direction
         if home_sensed then
            component_error = false
         else if pos_limit_sensed then
```

```
                component_error = true
        end
if component_error then
    begin
        notify astronomer of malfunction
        shut_down_system
    else
        continue_initialization
```

Using this example, a control system component error can be detected. Specifically, if the telescope is proceeding in a positive direction and passes the home position sensor without recognizing it, the telescope axis will eventually travel to its positive limit. As a result, one can conclude that a positive limit sense indicates the home positioning component is faulty. After the algorithm detects the unrecognized home position, a pop-up listing will alert the astronomer of the possible malfunction sources.

The above sample algorithm can localize component failures accurately; however, software developed by the diagnostic group at the NASA Ames Research Center can detect overall system performance and failure modes generated by the ADS-100 data points.

NASA Ames Research

AutoScope Corporation has been working with the NASA Ames Research Center (NASA ARC) on developing diagnostic software techniques using the ADS-100 unit as the vehicle to port test data.

One software package developed by NASA ARC is *DataHub*. It is a *Unix* based graphical interface which uses a highly flexible group of observation programs designed to facilitate real-time viewing of the ADS-100 data. The graphical interface allows visual access to all 37 TCS test points simultaneously. In essence, it is a 37 channel oscilloscope displaying each channel's allowable voltage range and instantaneous voltage level. Although this program is similar to the *CSVIEW* program developed by AutoScope, it has the advantage of residing on a *Unix* workstation which would front end the astronomical control system. These front ends allow other diagnostic software tools to access the ADS-100 data for further processing and severe number crunching.

To fully simulate the functional ability of the telescope control system, digraph models of the all possible control pathways of the TCS were developed by Christine Monahan of Recom Technologies at NASA ARC. The next passage discusses one of these -- the fault pathway for the telescope home sensors.

The previously discussed procedure for the home sensor algorithm used in *CSVIEW* can be seen below in a digraph representation. This modeling allows principal astronomers to draw more information from the system's functionality. It does not limit the user to simple component failures, but allows visualization of overall system failures. That is, you can foresee how a simple component failure will affect operation of the entire telescope control system. The following figure is the digraph example of the home sensors discussed earlier.

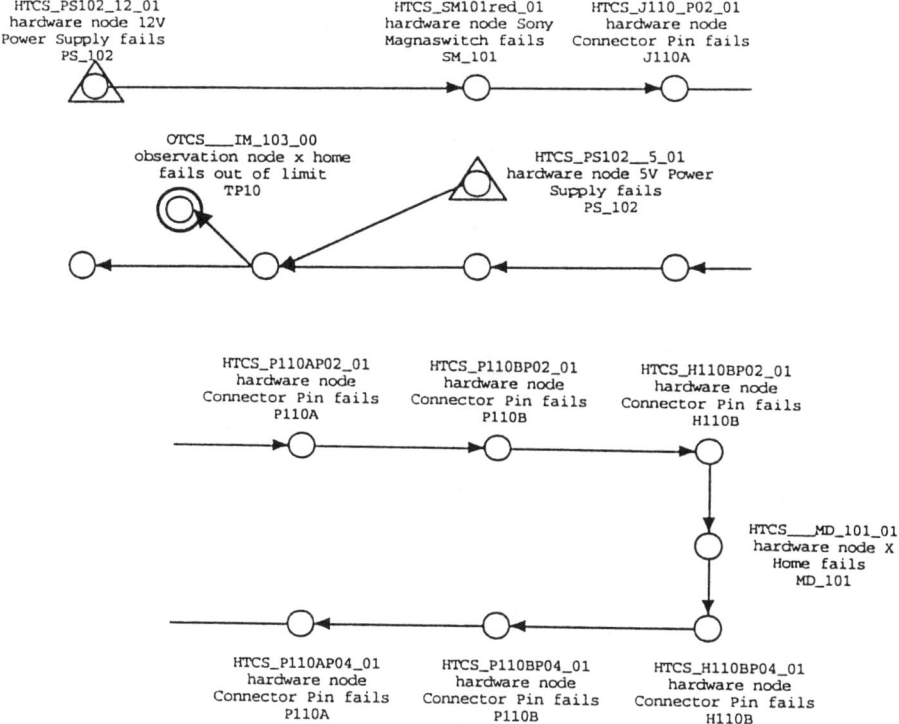

Figure 1. One of many telescope control system digraphs developed by NASA ARC. Shown here is the RA (Right Ascension) Home Sensor. Note that each node represents a hardware component within the TCS. The concentric circles labeled TP10 represent a TCS test point sampled by the ADS-100. Also, note that the digraph has been split down the middle due to paper and font constraints.

The digraph exhibits the interdependencies between the actual home sensor and its components. Figure 1 illustrates that successful home sensing of the telescope axis depends on several power supplies (PS_102_12_01 and

PS_102_5_01), a hardware port (J110A), an input module (IM_103), various cables (P1xxA and P1xxB), and the magnetic home sensor itself (MD_101). In the digraph each circular node represents a hardware component that can be a source for system malfunction. Each arrow "points" to a component that it immediately affects. That is, the control system's design principles dictate an immediate electrical dependency on adjoining nodes.

The concentric circles labeled TP10 represent an actual test point read directly from the TCS via the ADS-100. TP10's position was chosen to best diagnose the proper operation of the home sensor. If TP10 is not within its required parameter range, the home sensor digraph will back propagate towards the power supply PS_102_12_01. What this means is that anything which could affect proper operation of the magnetic home sensor will be highlighted and tagged as a possible source of error. These digraphs allow the telescope system operator to foresee all the components effected by a test point (TP10 in this case) that has exceeded its predefined boundaries. Ultimately, it predicts system failures due to a single component failure.

3.0 FUTURE CONSIDERATIONS

As mentioned earlier, automated diagnostics and health maintenance are important in the effort to realize truly remote operation of automatic telescopes. To reach this goal, the ATIS93 (Automatic Telescope Instruction Set 1993) standard will implement several new commands to accommodate key control system data, thus allowing nightly health monitoring of automatic telescopes.

ATIS was designed to be a standard interface protocol for requesting and receiving automatic telescope observations. The ATIS93 protocol may include a command set for the ADS-100 diagnostic unit. The command set would be terse in nature allowing test point data to simply be retrieved. For example, to access TP10 discussed earlier, the ATIS93 protocol would only need to know the telescope site, the control system module, and the test point to successfully retrieve the control system status. Other parameters may be added, such as length of time to monitor the test point and/or to sample several data points over a determined time interval.

ATIS93 will not only control unattended telescopes, but it will also have the ability to monitor the health of remote automatic telescopes.

4.0 SUMMARY

The complexity involved when coupling research telescopes and an automatic telescope control system introduces a need for automated health monitoring and diagnostic algorithms. Although automated monitoring of automatic telescopes may not replace on-site human engineers, it can reduce operational costs substantially for universities and institutions on smaller budgets. The ADS-100 is used as a vehicle to import diagnostic data such that smart digraph algorithms can be developed. These health monitoring algorithms localize possible system malfunctions, both in the control system and at the research telescope. Truly remote observations are possible by accessing key diagnostic test points via the ATIS93 standard while monitoring the telescope system performance. With this data, smart algorithms can maintain the health of the telescope and control system such that quality astronomical science is performed within the classroom.

ACKNOWLEDGMENTS

We gratefully acknowledge the financial support of NASA in the form of a Small Business Innovation Research (SBIR) Grant. We would like to personally thank Christine Monahan, Dr. Ann Patterson-Hine and all of the NASA Ames Research team for their exceptional insight and support.

REFERENCES

Eastham, P.C. 1993 in DataHub/TCSView User's Guide, NASA Ames Research Center Information Sciences Branch

Genet, R. M., and Hayes, D.S. 1989 in Robotic Observatories, AutoScope Corporation, Mesa, AZ.

Monahan, C.M., Patterson-Hine, F. A., and Iverson, D.L. 1994 in Automated Telescope Monitoring and Diagnosis, 106th Annual ASP Conference Proceedings.

Patterson-Hine, F.A., Boyd, M.A., and Iverson, D.L. 1992 in Automated Monitoring and Diagnosis of Telescope Control System, SALUTE Workshop Proceedings, NASA ARC.

Valentine, K. M.. 1993 in Self Diagnostics of Automatic Telescopes, 105th Annual ASP Conference Proceedings.

OPERATIONS ISSUES FOR THE HOBBY-EBERLY TELESCOPE

PHILLIP W. KELTON AND MARK E. CORNELL
McDonald Observatory, University of Texas, Austin, TX 78712

ABSTRACT   The Hobby-Eberly Telescope (HET) is now under construction at McDonald Observatory near Fort Davis, Texas. It is a joint project of the University of Texas at Austin, the Pennsylvania State University, Stanford University, the Ludwig Maximilian University in Munich, and the Georg August University in Goettingen. HET is an eight-meter class Arecibo style optical telescope whose primary mirror consists of an array of 91 hexagonal one-meter segments. HET will track objects with a multi-axis tracker at the top end of the telescope while the primary remains fixed at a 35 degree zenith angle. HET is a special purpose telescope optimized for spectroscopy, although imaging and other kinds of observing will be possible. Deliberate tradeoffs have been made between science capability and cost in an effort to maximize the performance/cost ratio of the telescope. One consequence is that the operational model for HET is unique and challenging, for scheduling and optimizing observations and the overall science efficiency. HET operations are scheduled to begin in 1997. This paper discusses HET operations issues and problems.

INTRODUCTION

The Hobby-Eberly Telescope (hereafter HET) is a joint project of the University of Texas at Austin, the Pennsylvania State University, Stanford University, the Ludwig Maximilian University in Munich, and the Georg August University in Goettingen. Operating policies will be established by the HET Board, which consists of representatives from the partner institutions. HET is now under construction at McDonald Observatory near Fort Davis, Texas. HET was previously called the Spectroscopic Survey Telescope (SST), and is scheduled to begin operation in 1997.

HET has the following key telescope attributes. See Figure 1 for a general view of the overall design. HET is an Arecibo-style optical telescope whose primary mirror will consist of 91 one-meter hexagonal Zerodur segments of spherical figure. HET will point at a fixed zenith angle of 35 degrees, and achieve sky coverage from -11 degrees to +71 degrees in declination by rotating the telescope in azimuth between observations. This will provide about 70% coverage of the sky observable from McDonald Observatory. The primary mirror will be fixed during observations, and tracking will be done in the spherical focal surface by a multi-axis tracking device. The tracker will have access to 12 degrees on the sky, and will consist of a large bridge beam implementing X and Y motions with an embedded hexapod assembly to carry the payload and implement additional

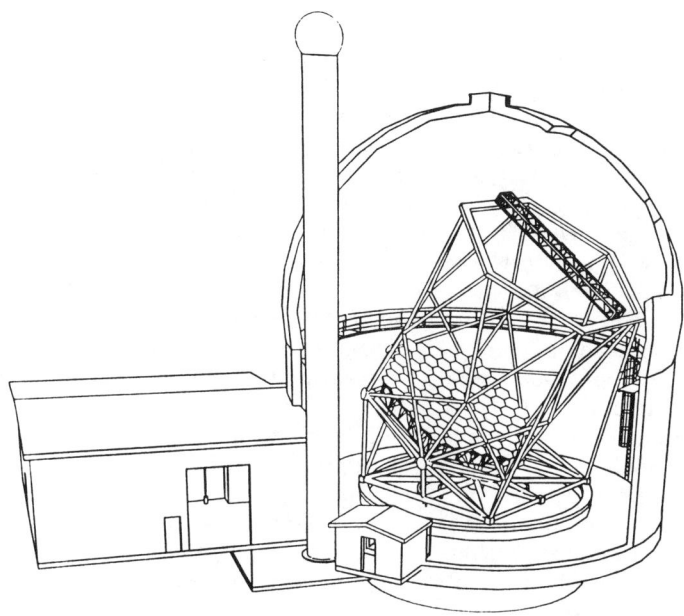

FIGURE I   Schematic of the HET design. Note the segmented primary, the 35 degree zenith angle, the top end tracker, and the CCAS tower. The main instrument room is underneath the telescope base.

axes of motion. The payload will include a baffled spherical aberration corrector, acquire and guide modules, prime focus instrumentation, and the mechanism to feed the signal via fiber to non-prime focus instruments located below the base of the telescope. A separate tower will house the Center of Curvature Alignment Sensor (CCAS) instrument, toward which the primary will be pointed during alignment procedures. This unique design leads to many challenges in scheduling, implementing the top end control, and achieving efficient operation.

Recent descriptions of the HET telescope are given by Ramsey et al (1994) and Sebring et al (1994). HET is designed to implement deliberate tradeoffs, in which some limitations in the scientific capability are accepted in return for the elimination of major cost drivers. HET will be used primarily for spectroscopy and is also well suited for narrow field imaging. It is a special purpose rather than a general purpose telescope. Some compromises have been made (e.g., in sky coverage, tracker access, and vignetting), but the overall goal is to maximize the performance/price ratio in pursuit of those astronomical research objectives for which HET is designed.

Spectroscopy is the primary HET scientific mission and this will be reflected in the initial instrumentation. The baseline instrumentation plan calls for a fiber-fed medium resolution spectrograph as the first light instrument. Near term instruments will include a fiber-fed high resolution spectrograph and a prime focus instrument that will combine low resolution spectroscopy and imaging. Near infrared instruments are in the far term plan.

HET OPERATIONS: BASIC REQUIREMENTS AND GOALS

HET will be operated for benefit of the partnership by McDonald Observatory, through an on-site HET staff that reports to the McDonald Observatory Superintendent. HET requirements call for queue-scheduled observing to be the primary mode of operation over the long term, with astronomer-on-site dedicated mode observing also available as a secondary mode of operation. One essential HET requirement will be to ensure that the prescribed overall time shares are maintained for each partner institution. In queue-scheduled mode, observations will be interleaved among the partners to achieve maximum efficiency, such that each partner receives its proper amount of time when summed over an interval of time that is still to be determined. Only "clear hours" will be counted against each partner's allocation, except when the telescope is assigned in dedicated rather than queue-scheduled mode. However, the issues of how to define "clear hours", how to account for "good seeing" time versus "bad seeing" time, and how to factor in quality of acquired data in the accounting of each partner's time allocation are still open questions.

The overall procedure for observation planning, proposal handling, and scheduling will be implemented over the computer network with interactive support from suitable software tools. Preliminary inspection of the data before final delivery should also be supported via the network, from distant places such as Austin, Pennsylvania, California, and Germany. Data delivery will be through magnetic, electronic, or other machine-readable form, accompanied by sufficient information about the telescope and instrument to allow full calibration and reduction of the data for publication. HET observing will not require the presence of the astronomer once routine queue-scheduled operation is achieved. For queue-scheduled observing, questions of risk and data verification come up for observations made by on-site operations staff when the proposing astronomer is not present. The need to mitigate such risk leads to a requirement for a good mechanism that allows astronomers to inspect their preliminary data through the computer network and provide feedback on data quality into the overall observing procedure.

Another basic HET requirement will be to minimize inefficiencies from telescope and instrumentation configuration changes, and to minimize overhead and lost photons whenever possible. Intelligently automating the telescope will be crucial in this regard. The HET has many potential ways to incur excessive overhead and lost throughput, probably more so than in typical conventional telescopes. Significant loss of photons to the science exposures can occur, for example, through increased image size due to primary mirror segment misalignment, excess time lost to primary realignment or instrument calibration, excess azimuth and/or tracker slews, dead time during schedule gaps, or losses of light due to incomplete illumination of the primary whenever exposures cannot be scheduled in the central region of the tracker.

HET AS A "ROBOTIC TELESCOPE"

There are many examples of robotic telescopes. Telescopes in space offer perhaps the ultimate example, and many radio telescopes have long been known for either automatic or remote operation. Automatic Photometric Telescopes (APTs) that do photoelectric photometry are now in widespread use, and automatic imaging telescopes have also been implemented.

Automatic spectroscopic telescopes are not common, however, although various groups have indicated their desire to build them. HET will specialize in spectroscopy, though not exclusively, and will be one of the first major telescopes to seek automatic operation in spectroscopic observing. HET will seek automatic operation by single operators, but not fully unattended robotic or remote observing. The HET operations model is certainly unique, though it does share many common problems with other non-spectroscopic robotic telescopes such as those mentioned above.

HET's robotic nature is largely due to the multi-axis top end tracking subsystem, the segmented primary, and the requirement to operate in queue-scheduled mode. There are about 300 total degrees of freedom in HET, including 273 for the primary mirror segment control (3 actuators for each of 91 segments). Coping with all of these degrees of freedom is a major operations issue for HET. The tracker is essentially a sophisticated robot in itself, with ten degrees of freedom in the baseline design. Logically there are six axes in the tracker: X and Y for the large motions, Z (primarily for focus), theta and phi (angular motions to keep the payload normal to the primary at all times), and rho (position angle of the instrument package). Physically there are ten axes of motion in the current preliminary tracker design: X1 and X2 for the two ends of the large bridge beam, Y to transport the payload along the bridge beam, H1 through H6 for the six linear elements of the Stewart platform device (a "hexapod") which will carry the payload itself, and rho for the rotation angle. There will be additional axes of motion in the payload itself, including the spherical aberration corrector, acquisition and guide subsystem, fiber feed mechanism, and prime focus instrumentation.

Evolving to automatic "robotic" operation is a long term HET goal, and we clearly recognize that a significant learning process will be necessary.

HET OBSERVATION DEFINITION

The fixed altitude primary of HET will point at a 35 degree zenith distance. This fact leads to the requirement to track objects in the spherical focal surface according to the basic mode of operation now described. The scheduling and acquisition of targets is unique for HET. Once objects are acquired, however, HET observations will be fairly typical of spectroscopic observations on other large telescopes, within the constraints of the limited HET observing windows.

Assume that the scheduling algorithm has already selected the next object to be observed. Assume it has also determined the time window during which the tracker can intercept it and the azimuth for the particular declination of the object in order to optimize its trajectory through the tracker. If the current telescope azimuth is not sufficiently close to that required to observe the selected object, then the telescope must be moved in azimuth. The motions during an azimuth change must be sufficiently smooth that realignment of the primary mirror is not needed afterward. Grouping observations together to minimize overhead from azimuth slewing and possible mirror realignment is highly desirable but can only be done within the constraints of the scientific priorities and the ability of the scheduling algorithm to make reasonably optimum schedules. Even if the selected object will transit the tracker at the current azimuth setting, a small change in azimuth may still be warranted to optimize the exposure's exact trajectory, transit time, and aperture function. The aperture function is the amount of the primary mirror's diameter that is seen by the payload as it transits the tracker, which varies from about 6.5 meters of primary mirror diameter near the tracker's edge to about 9 meters in the central range of the tracker.

Each target's predicted trajectory will be downloaded to the tracker servo subsystem for acquisition and tracking. The transit time defines the maximum exposure time, which varies from about 45 minutes in the south to near 2 hours in the north. If light losses can be significantly reduced by making a small adjustment to the azimuth setting before beginning a science exposure in order to optimize the aperture function, then the tradeoff between increasing the effective aperture and consuming some time in azimuth resetting and occasional extra realignment must be resolved. The bottom line is that there is a premium on observing in the tracker's central region whenever it is feasible to do so.

The selected object will be acquired by intercepting it at a precisely timed moment in the tracker field and switching from acquisition to tracking mode. The tracker servo subsystem will track the downloaded trajectory by implementing for all of the tracker axes the velocities and accelerations necessary to follow the positions defined by the trajectory. Meanwhile, the object is identified and verified in the CCD acquisition camera (or by some field recognition algorithm), centered on the fiber or other entrance aperture of the instrument, and two or more guide stars are selected and acquired by the guide CCDs. The science exposure is then made while the autoguider determines error signals and passes them to the tracker servo to close the overall tracking loop. The guider will not only correct for errors in right ascension and declination but also in focus. The data are then annotated, displayed, and stored. Any specified calibration frames are also taken and stored, and the system log is updated in the observation data base. The final step is to verify data quality and the completeness of the observation set. Exactly when to take the calibration frames and how and by whom the data quality is verified are operational problems yet to be addressed.

Quality control is very important for telescopes operating in automatic mode. Automated execution of queue-based schedules is one of the long term HET goals. System tests or other quality control checks will have to be interleaved with science observations to maintain a satisfactory level of data quality. Details on how to achieve adequate quality control have been demonstrated for APT telescopes doing photometry, but are yet to be determined for automatic spectroscopic telescopes such as HET. Gain and read noise of the CCDs are easy to check, but other checks such as signal-to-noise ratio of science spectra and focus across the spectrum are not easy to automate.

A data base of authorized observations defining necessary observing information will be central to the HET observing process. Updating this data base as observing proceeds will be fundamental to HET operations, and data base design will also be important. Defining complete observations (object plus calibration data) can be complex, and doing so to the extent needed to support full queue scheduling and automatic operation of HET will represent a cultural change on the part of astronomers used to conventional modes of observing. HET will not be too different from the Hubble Space Telescope in this regard. Implementing good software tools to simplify the observation planning and definition process for HET observers will be important.

HET observation definition data base requirements include the following general categories. Housekeeping entries include proposer institution and observer names, object name, and observation number. Object details include coordinates, proper motions, magnitudes, and orbit or ephemeris information for non-sidereally moving objects. The data base should include guide objects and their coordinates, magnitudes, and proper motions if available. The HST Guide Star Catalog is one obvious source for guide star information, although its faint limit is too bright for some HET observations and a significant impact on the operations model can be expected for those cases. Priorities defined in the data base should include both observer-defined and TAC-defined, and a tally of actual accrued observation time by partner percentages will also need to be maintained in the data base as observing proceeds. The observation data base must also include sufficient definition of exposure requirements to guide the scheduler and the operator in queue-scheduled mode, including nominal exposure time, signal-to-noise ratio requirement, seeing and transparency requirement, sky background requirement, and number and timing if a series of observations is being defined. Instrument configuration is another category, including grating, angle, order, fiber or slit position angle, or other parameters for spectroscopic observations. Finally, calibration needs to be defined include flats, arcs, darks, bias frames, and standard stars. The flats, darks, and bias frame requirements will standardize as we learn more about the HET instruments.

HET SCHEDULING: PROBLEM STATEMENT AND GOALS

The unique nature of HET implies that a sophisticated scheduling subsystem will be needed to use available telescope time efficiently. The goal of HET scheduling procedures and software will be to maximize the efficiency of telescope usage while minimizing changes of telescope and instrument configuration, consistent with the scientific goals as set by the telescope time allocation committee, observer-defined priorities, and the mandated division of telescope time among partner institutions. The scheduling algorithm will seek to minimize "dead time" during a routine night of observing, and will also attempt to minimize the loss of photons by scheduling observations as near the center of the tracker field as possible in order to maximize the aperture function for each observation. A parallel goal will be to minimize overhead due to realignment of the primary mirror and excess tracker and azimuth drive slewing.

HET SCHEDULING: SOME OPTIONS TO CONSIDER

We are now studying various options and tradeoffs for implementing the HET scheduling subsystem. We will begin by implementing an "interactive planning tool" program that will incorporate and visualize the basic HET geometry and graphically show proposed observations relative to the telescope azimuth, tracker orientation, date and time, exposure time limitations, potential schedules, and so on. This tool would support the manual generation of rough, non-optimized schedules and is described more in a later section. Beyond this basic tool, we recognize that writing all of the scheduling code ourselves "from scratch" would be a complex and expensive task to be avoided if at all possible. We have studied specific rules that might be devised to govern HET scheduling, and believe that implementing a full scheduling program in-house based on such rules is not the way to go. Furthermore, existing planning and scheduling software developed for other telescopes may be adaptable to our problem, and so we are now considering available alternatives.

We are interested in studying all potentially applicable models. Some of the planning, proposal processing, and scheduling software developed at STScI and used for HST could potentially be applied to HET. The best known of these programs is SPIKE (Johnston 1992), an AI-based system developed to augment the original HST scheduling software. Although SPIKE has many useful attributes for generating schedules, it may not support flexible scheduling in real time to the extent that HET needs; this is still an open question, but insufficient support for flexible scheduling would be a severe limitation from our point of view. Another AI-based system is the CERES system developed at NASA Ames (Bresina et al 1994). The CERES system implements a concept called the "Associate Principal Astronomer" (APA), in which the software acts as an assistant to support the activities of a Principal Astronomer (i.e., a real person who is in charge of a remotely operated telescope). This system is being applied initially to automatic photometric telescopes (APTs) and has potential to be adapted for HET use. The BAIT telescope (Berkeley Automatic Imaging Telescope, Richmond et al 1993) is another interesting model that implements algorithms and techniques which might be adapted to HET use. Other large telescopes recently constructed or now being built (e.g., WIYN and Gemini) may also implement approaches to scheduling which are of interest to HET.

These various models employ different approaches to describing observation details, incorporating them in their data bases, and sequencing resulting telescope control functions. Although SPIKE is Lisp-based, it makes use of front-end tools for electronic proposal submission that are ASCII files in a keyword-value format. Those tools are augmented by validation tools that check for syntax errors, the goal being to have automatic transformation of proposal data to the actual planning and scheduling parameters. The APT telescopes introduced the use of ATIS (Automatic Telescope Instruction Set), and the NASA Ames system currently employs the 1993 version called ATIS93 (Pyper 1994) to describe photometric observations. Whether it is reasonable to use an extension of ATIS93 to describe spectroscopic observations for HET is an open question. Another alternative with some appeal would be to adopt a FITS-header based style as in the BAIT system. Which particular technique is best suited to HET is an important question that needs more study before finalizing our design.

The best strategy for HET is adoption of a modular approach that allows for flexibility and that does not prematurely lock us into any particular observation definition language or command language for sequencing telescope operations under queue-scheduled automatic operation. Implementation should avoid hardwiring any particular observation definition language directly into the code. Use of a modular approach will allow adoption of different external (i.e., to observers) and internal (i.e., to the run time system) languages for defining and executing observations, with modules to make necessary format transformations. Use of unique serial numbers to tag each proposed observation set so they can be accurately tracked through the system is desirable. Combining planning and scheduling tools from one or more external sources such as those mentioned above with HET-specific tools developed in-house may be the best approach. Careful attention to modular design will allow for incremental development and replacement of specific elements of the HET planning and scheduling system, and also support whatever evolution is needed in the software as we use HET and learn from our experience.

HET SCHEDULING ISSUES

Many issues arise when thinking about algorithm-based scheduling of HET observations. The term macroscheduling describes the general process of assigning observations on a timeline with very coarse resolution, for example on the order of days or weeks. The term microscheduling, called loading by some, describes the process used to assign observing events to a nightly or other shorter time scale; in the case of HET, an increasingly detailed model of HET operational constraints would be applied to produce schedule predictions which are reasonably accurate if and when executed. It is not necessary that either process produce truly optimized schedules, since the likelihood that events in real time will interfere with a continuous and precise execution over a full night is fairly high. Schedules are called "broken" when that happens, and dealing with this case in an effective way will be important when automatic queue-scheduled operation becomes routine for HET.

There are many conditions under which broken HET schedules could occur. One such condition would be a lack of suitable guide stars. Another would be the presence of changing conditions (e.g., weather, seeing, transparency, or onset of misalignment). These different conditions will all be difficult to deal with in practice, and how to effectively adjust schedules in real time is a serious open question. For example, if the seeing or transparency degrades significantly, then the resulting loss of light into the spectrograph during a long exposure would make necessary an increased exposure time to compensate, and the remaining sections of a carefully pre-computed and optimized schedule could be instantly invalidated. We expect to be taking long exposures often, however, in which case increasing the exposure time may not be a realistic option given the HET design. Deciding when a spectroscopic HET observation is complete may not always be easy even in the absence of such changing conditions, since deciding when signal-to-noise ratio requirements have been met will sometimes be difficult in practice when no astronomers are present. This is why a network-based mechanism for proposing scientists to interact with preliminary data is important.

Complications like these lead to a requirement that HET queue-scheduled operation be able to support flexible scheduling. This means that the algorithm must allow rapid recalculation of modified schedules whenever "broken schedules" develop. The observation model and scheduling algorithm must deal with the entire sequence of an observation, not just the science exposure within the full observation set. The practical problems associated with interleaving multiple instruments and calibrations and coping with uncertainties in how long it really takes to achieve target acquisition and guide star setup must all be solved efficiently in any automated execution mode.

Numerous other factors will complicate HET scheduling. These include matching observations to their sky brightness requirements, which is becoming increasingly important as high resolution spectroscopic observers go to increasingly fainter objects; representing and executing time-dependent aspects of observation requests, e.g., for variable stars; defining and implementing the rules for dealing with "sludge" which will develop in the observation data base, i.e., lower priority observations which never rise in priority enough to actually be observed should be pruned periodically; determining if it will really be practical to interleave observations between different instruments to optimally match observing conditions to instrumental performance and program requirements; and determining if the underlying mechanical repeatability and instrument stability are sufficient so that all components of a "complete observation" do not have to be obtained before any configuration changes can be allowed to occur.

Queue-based scheduling and execution for HET will be a long learning process characterized by increasingly more sophisticated and efficient algorithms. We plan to assign blocks of time during HET commissioning to individual astronomers to use the telescope in manual mode so that their learning experiences can be fed back into the queue scheduling algorithm as it evolves. Developing a software tool to simulate execution of schedules should also prove useful.

HET INTERACTIVE PLANNING TOOL

An Interactive Planning Tool (IPT) incorporating HET-specific details will be needed to support HET operations. This software will support interactive planning and observing without any pretense to achieving optimized use of the HET capabilities. Over the long term, we plan to integrate it with additional planning, scheduling, and execution software tools in a modular fashion so that we evolve toward a combination of in-house HET-specific code and external software that collectively implements the HET scheduling subsystem.

IPT needs to implement a variety of functions. It must manage a data base of observation requests, and add whatever HET-specific details are needed to transform user supplied information into complete observation definitions. It should allow generalized input of time, azimuth, and object coordinate information, either manually or from object data base files or catalogs. It should be able to compute object trajectories within the tracker field, compute the aperture function for specific azimuth and object combinations, and display the results graphically for the observer or operator. IPT should contain a model time allocation of observations, including target acquisition with azimuth changes if necessary, guide star setup, and overhead due to readout, storage of data, and other components of a complete observation. Using all of this information IPT should be able to compute optimum azimuth settings and time windows for specific observations. If exposure meters are built into instruments, however, it is not clear how to incorporate the associated exposure time uncertainty into the IPT time allocation model. A signal-to-noise ratio predictor and/or empirical exposure time predictor that can suggest exposure times for available combinations of instrument, spectral resolution, bandpass, and so on would also be an essential component of IPT. It would assist in planning observations as well as guiding the process of creating schedules that make a reasonable allocation of available time within the tracker field of view. Finally, IPT must produce output files of possible observation schedules for use directly at the telescope or for other analysis.

We have implemented a limited prototype version of IPT with a subset of this functionality as part of phase one of a design study of HET queue scheduling. The initial design study goals were to simulate queue-based scheduling, study rules that might apply, formulate strawman object data base requirements, and gain experience before attempting to define the master HET scheduling procedure. A sample object data base of several thousand objects was generated in a strawman HET format by ten potential HET observers for use in scheduling simulations and testing prototype concepts. Conducting simulations will also help us to evaluate how our in-house HET-specific code can be combined with other planning and scheduling software from external sources before prematurely committing to any inadequately understood long term solution.

OTHER HET OPERATIONS ISSUES

HET has other important operations issues besides the unique tracking and scheduling related issues that have been emphasized thus far. Servicing the top end will be difficult because HET has no conventional service position due to its fixed 35 degree zenith angle and also because HET will not have mirror covers (e.g., filling dewars, changing instruments, protecting the primary mirror, or troubleshooting problems). How data quality will be validated in the absence of on-site observers by telescope observer/operators is another important issue. Data delivery and security under queue-scheduled operation will also require some attention to detail. Minimizing the amount of night time devoted to calibration exposures is another challenge with the potential for interleaved operation, multiple instrument usage, and generality of scheduling that HET will ultimately allow. Another operations issue involves deciding which on-site personnel will be required, how many, what level of training is necessary, and so on. All new telescopes face such decisions, but the unique operations model planned for HET over the long term makes these issues particularly important. Finally, we wish to support limited access to HET through Internet for engineering staff to troubleshoot and debug problems and monitor system performance remotely, and some planning for this is essential.

HET EFFICIENCY ISSUES

A major goal of automated operation will be to increase the efficiency of HET observing. Just as one goal in the telescope and instrumentation is to maximize the optical throughput through careful design and implementation, an analogous goal in the operational software will be to limit time lost to inefficiencies in usage and seek to maximize the amount of time spent collecting science data. Observations with high TAC ranking would, however, override such considerations, e.g., for an event like the July 1994 comet impact on Jupiter. Efficiency is important for any telescope. With HET, however, the efficiency issue is especially important since the telescope's unique nature implies that the learning curve toward achieving efficient operation may be steeper than usual.

A good framework for discussing efficiency of operation is to partition the observing process after Petro et al (1991). This allows one to measure efficiency by considering the percentages of time actually spent in science data acquisition versus other related but necessary activities. "General setup" includes focus, alignment, etc. "Instrument setup and control" involves configuring instruments to the specific requirements for each science observation. "Target acquisition" includes pointing, target verification, and centering in the aperture. "Guide star acquisition" includes not only finding guide stars but also efficient setup of autoguiders. "Exposure" includes not only the science object but all required calibration exposures as well, including flats, darks, wavelength calibration exposures, and standard stars. "Data recording" includes generation of FITS headers and keywords, annotation of observations, and the time to write and verify data files. "Overhead" includes slewing, time when no requested observations can be easily obtained, and other unscheduled down time. "Other" includes time when no science object is being observed but other useful work can be done, such as pointing or other essential calibrations.

Various telescopes have been analyzed in this context to compute observing efficiency and to identify possible improvements (e.g., HST, IUE, and NTT; see Benvenuti 1992). With HET's unique mode of operation, it will be especially important to keep these considerations in mind as the telescope is built and put into operation. HET observations will be fairly conventional in principle but more complicated in practice than a conventional telescope due to constraints that follow from the telescope design itself. For example, the tracker operation and the aperture function will have a strong bearing on the overall throughput of the HET telescope system.

Many factors will require careful implementation and attention to detail to achieve efficient operation with HET. Mirror realignment dead time is required to consume less than 10% of available observing time, although the long term goal is to do much better than this and develop a focal surface based technique to supplement the Center of Curvature Alignment Sensor technique. Optimizing the pointing/setup/exposure/calibration cycle within the tracker is another important goal. Pointing accuracy will determine the extent to which finder charts or their electronic equivalents are needed, and how much observing time is consumed in locating and verifying objects and/or identifying known positions and executing offsets. Implementation of flexible scheduling in response to changing conditions, achieving concurrency of operation wherever possible, implementing efficient command grouping (e.g., macros, scripts, command files), and achieving efficient initialization and restart will be other important HET efficiency considerations. Using instrument configuration files to reconstruct setups from header keywords is another useful technique, although in practice it is more likely that we will generate the headers first and then do the instrument setup from the preset parameters.

McDonald Observatory's experience operating large telescopes suggests that there are many subtle influences that may be hard to enumerate or quantify yet which strongly affect the efficiency and overall performance of large telescopes. This discussion illustrates some of these issues. As HET implementation proceeds toward the operations phase, it will be essential to carefully address these efficiency issues.

STATUS

HET is a unique telescope with a challenging operations model to implement. The schedule calls for the building, dome, telescope structure, and primary mirror truss to be finished in mid-1995. Installation of the optics, tracker, primary mirror control system, and other subsystems is scheduled for later in 1995 and throughout 1996. The current software effort is focused on defining the overall system architecture and implementing basic control of the hardware. Operations issues including scheduling are now under study, and implementation of operations support software will accelerate in 1995 after further definition of requirements and study of concepts, alternatives, tradeoffs, and costs.

ACKNOWLEDGMENTS

Thanks are due to Randy Ricklefs and Jim Foster of the McDonald Computing Services Group, who have contributed to the effort to define HET operations support software, and to Tom Sebring (HET Project Manager), Larry Ramsey (HET Project Scientist), and the entire HET Project Team. Tom Barnes is leading the McDonald Observatory effort to plan the transition from HET construction to full HET operations.

REFERENCES

Benvenuti, P. 1992, ST-ECF Newsletter, February 1992, 26

Bresina, J., Drummond, M., Swanson, K., & Edgington, W. 1994, A. S. P. Conf. Series, volume 55, D. M. Pyper and R. J. Angione, editors, 216

Johnston, M. D. 1992, AAAI Spring Symposium Series "Practical Approaches to Scheduling and Planning," Stanford, CA

Petro, L., Stockman, P., & Whitmore, B. 1991, STScI Newsletter, November 1991, 4

Pyper, D. M. 1994, A. S. P. Conf. Series, volume 55, D. M. Pyper and R. J. Angione, editors, 165

Ramsey, L. W., Sebring, T. A., & Sneden, C. A. 1994, SPIE Technical Conference 2199, Advanced Technology Optical Telescopes V, L. M. Stepp editor, Kona, HI, 31

Richmond, M. W., Treffers, R. R., & Filippenko, A. V. 1993, Publications A. S. P., 105, 1164

Sebring, T. A., Booth, J. A., Good, J. M., Krabbendam, V. L., Ray, F. B., & Ramsey, L. W. 1994, SPIE Technical Conference 2199, Advanced Technology Optical Telescopes V, L. M. Stepp editor, Kona, HI, 565

Automating Mission Scheduling for Space-Based Observatories

Nicola Muscettola
Recom Technologies
NASA Ames, MS: 269-2
Moffett Field, CA 94035-1000

Barney Pell
RIACS
NASA Ames, MS: 269-2
Moffett Field, CA 94035-1000

Othar Hansson
Heuristicrats Research, Inc.
1678 Shattuck Ave. Suite 310
Berkeley, CA 94709-1631

Sunil Mohan
Recom Technologies
NASA Ames, MS: 269-2
Moffett Field, CA 94035-1000

Abstract.
In this paper we describe the use of our planning and scheduling framework, HSTS, to reduce the complexity of science mission planning. This work is part of an overall project to enable a small team of scientists to control the operations of a spacecraft. The present process is highly labor intensive. Users (scientists and operators) rely on a non-codified understanding of the different spacecraft subsystems and of their operating constraints. They use a variety of software tools to support their decision making process. This paper considers the types of decision making that need to be supported/automated, the nature of the domain constraints and the capabilities needed to address them successfully, and the nature of external software systems with which the core planning/scheduling engine needs to interact. HSTS has been applied to science scheduling for EUVE and Cassini and is being adapted to support autonomous spacecraft operations in the New Millennium initiative.

1. Introduction

The use of spacecraft observatories has dramatically expanded the horizons of astronomy and profoundly influenced our understanding of the Universe. Operating such observatories has always posed challenges that are not usually present in the operation of ground based telescopes. Once in space, instruments are unaccessible for maintenance and repair. Therefore, safe operation of the space facility has to be insured to a very high degree. Streams of commands executed by the spacecraft must be guaranteed not to put the spacecraft in any dangerous situation (e.g., pointing an instrument directly toward the Sun or having too many subsystems on at the same time draining electric power from internal supplies). Moreover, space observatories are unique and precious resources for the scientific community. Very often the requests for observations exceed what can be possibly accomplished during the limited lifetime of a mission. Accommodating a high number of requests while satisfying the observation constraints calls for the advance generation of an efficient observation schedule.

Because of the combination of safety and efficiency issues, automating the generation of detailed sequences of spacecraft commands is a challenging computer science problem. Its solution will facilitate spacecraft operations, reduce operational costs and ultimately allow scientists to operate spacecraft more directly, with potentially high scientific payoffs.

In this paper we describe the HSTS (Heuristic Scheduling Testbed System) project (Muscettola 1994a). The project is developing technologies for automated planning and scheduling and is applying them to the automated generation of spacecraft commands. We first give an overview of science mission operations with particular emphasis on planning and scheduling issues and we look at the main cost drivers. On the basis of this discussion, we point out several requirements on practical solutions to the computational problems in this domain. We then discuss the HSTS technology and our progress to date in providing a solution which satisfies these constraints.

Mission operations have traditionally accounted for a large part (40% on average) of the cost of a mission. Examples of such activities are science scheduling, command up-link (i.e., the generation and communication to the spacecraft of "correct" command sequences), and the monitoring of spacecraft operations. Operations teams vary in number depending on the mission, but range from tens of people for a small astrophysical observatory (e.g., EUVE) to several hundreds in the case of larger missions (e.g., Voyager, Hubble Space Telescope, and Cassini).

At present, the space program is exploring ways to reduce the cost of its space missions. One avenue is to use several small spacecraft in order to achieve the goals achievable by a single large spacecraft. This approach has the advantages of reducing the complexity of spacecraft operations and reducing spacecraft development costs through the reuse of standardized hardware components. However, when the cost of software development is taken into account, this approach will not necessarily be effective. The structure and scientific tasks of these smaller spacecraft are likely to vary greatly from one mission to another. Without software reuse and cross-fertilization across missions, mission support systems will be developed separately, as has been the case with previous missions. If this approach is followed in the future, we will encounter an even bigger operations problem than at present.

One of the principal activities in mission operations is the generation of sequences of commands for the spacecraft. This involves a fairly complex process with several distinct stages. Proposals are submitted to the organization responsible for managing the mission's scientific activities. A proposal is typically organized in a program of observations all aiming to achieve a scientific objective. Each observation in a program can specify the use of one of several instruments on the spacecraft. A program can contain a diverse set of temporal constraints including precedences, windows of opportunity for groups of observations, minimum and maximum temporal separations, and coordinated parallel observations with different viewing instruments. Proposals are reviewed on the basis of their scientific merit and a set of observations is accepted and selected for execution by the spacecraft.

Several missions have adopted a process that generates a command sequence in two separate steps: long-term scheduling and short-term scheduling.

Long-term scheduling takes into account an aggregate characterization of the spacecraft's observation capacity and long term periodic viewing constraints. In the case of Earth orbiting observatories, for example, a periodic viewing constraint arises because of the change of the position of the Sun with respect to a celestial object during a year. When a celestial object falls within a given angle centered around the Sun's position, it is not possible to observe the target because this would require pointing instruments too close to the Sun and therefore endangering the spacecraft and instruments. When the object is farther away from the sun, observations are possible but not preferable because of the possibility of sunlight polluting the instrument. Taking into account such constraints, long-term scheduling algorithms (Bresina 1994; Johnston & Miller 1994) distribute the accepted proposals on the mission timeline, decompose the time-line in regular intervals (e.g., monthly or weekly) and determine which set of observations need to be executed for each interval.

Short-term scheduling takes the result of long-term scheduling, focuses on one time interval and the associated observations, and generates the sequence of commands that the spacecraft will execute to implement the observations. This requires the consideration of a much more detailed characterization of the spacecraft subsystems and operations constraints. For example, most targets observed by satellites in low-earth orbit are periodically occluded by Earth (once per orbit). Similarly, interplanetary probes must synchronize their communication to Earth during the periods of visibility of the Deep Space Network antennas (which depend on the daily rotation of the Earth). Short-term scheduling must also take into account several stringent constraints on the operations of spacecraft subsystems. These include limited available electric power, and maintenance of acceptable temperature profiles on the spacecraft structure. The pointing subsystem is responsible for orienting the spacecraft toward a target, and locking it in a well defined position of the field of view of the designated scientific instrument. Spacecraft with several instruments may not have enough available power to keep all instruments operational simultaneously. Moving an instrument between operation and quiescence may require complex reconfiguration sequences which must be coordinated among various instrument components.

Reconfigurations must also be appropriately synchronized among different instruments. Data can be read from the instruments and directly communicated to Earth; more usually it is temporarily stored on an on-board memory unit and communicated to Earth at a later time.

The complexity of the short-term scheduling problem and the large variety of constraints that need to be considered from one mission to the other makes it difficult to devise totally automated solutions to the problem. So far, short-term scheduling is a process that requires various degrees of human involvement. In the development and application of the HSTS technology we aim to demonstrate that Artificial Intelligence (AI) planning and scheduling can provide the basis for a more systematic and cost-effective approach to the development of automatic and mixed initiative mission operations systems, in particular for short-term scheduling. To be successful, however, we have not approached the problem by providing a turn-key, totally automatic system that is not integrated into the current, mostly manual scheduling process. A proposed framework may not be useful (and therefore not be used) if it tries to duplicate functionalities already provided, if its problem solving and representation paradigm does not easily accommodate the information typically used by human operators, or if it does not enable a mixed-initiative problem solving process between human and computer.

The rest of this paper is structured as follows. Section 2. describes the organizational process of building a science schedule and its computational challenges. Section 3. provides a set of necessary features for a planning and scheduling tool to be practical in supporting this process. Section 4. describes the architecture of HSTS. In particular, we elaborate on desiderata for one particularly important requirement, that of an appropriate language in which to model spacecraft components and scientific objectives. This discussion is grounded in our modeling language, HSTS-DDL, for use with the HSTS integrated planning and scheduling system (Muscettola 1994a). We then discusses the integration of external knowledge within our modeling language and scheduling tool. Section 5. discusses current and future applications of HSTS.

2. Science Schedule Construction

A fully automated science mission scheduler would take as input a prioritized set of basic science goals and constraints (such as "observe the impact of comet x on planet y" and "never point the instrument at the sun"), and would produce as output a detailed list of primitive spacecraft instructions (a *sequence*). The execution of a sequence would lead to the achievement of the science goals subject to the constraints. Sequence quality can be measured in terms of the number and importance of the science goals achieved, in terms of the efficiency of use of available resources, and in terms of the effort needed to validate that the sequence satisfies its constraints.

There is an important tradeoff in sequence generation between sequence optimality and tractability. Micro-optimized sequences (i.e., sequences optimized at the level of individual spacecraft-instructions) may achieve more science goals, but generating such optimized sequences can be very expensive. This is in fact the traditional approach which has been used, for example, in the Voyager missions. Voyager was a fly-by mission in which each period of a few months of intense active observation was separated by years of relative inactivity. During the Neptune encounter, for example, about 1000 observations were executed over a period of about 5 months. In order to integrate *observation requests* from about 50 scientists into a sequence, a team of approximately 50 *support schedulers* worked full-time for a period of 4 years in advance of the encounter. The problem was an order of magnitude larger for each of the encounters with the closer planets of Jupiter and Saturn. Voyager had 11 science instruments that could operate in parallel. Each support scheduler approached the sequencing of each observation request on each instrument as a micro-optimization problem, and further optimization was then achieved by optimizing combinations of observation-specific sequences.[1]

This extremely labor-intensive approach to sequencing is no longer viable. This is due in part to limitations on available budgets and in part to the nature and projected number of future missions. Thus, most space missions have abandoned the micro-optimization approach used in Voyager and have adopted an organizational model for science scheduling that emphasizes reuse of a small set of standardized sequence components.

Before any spacecraft sequencing is done, a set of scientists and engineers first develop a small set of programs composed out of primitive spacecraft and instrument instructions, which are intended to achieve certain common subgoals. Scientists and engineers refer to these programs as *modules*. For example,[2] one module might lock the spacecraft onto a target once the instrument is pointing in the vicinity of the target; another module might turn on the tape recorder; another module might dump the contents of a specified tape recorder and send the results back to earth; and yet another module might move the spacecraft's field of view from one location to another, an operation called *slewing*. It is important to note that the modules themselves are not necessarily fixed action sequences, but may be conditional and parameterized networks of commands. This means that substantial work may be invested to verify that calling a particular module with a given set of parameter values in a particular spacecraft state will have its intended consequences. Of course, as their name suggests, the modules are designed to be modular and, thus, relatively insensitive to external conditions.

Once a set of modules has been defined, the iterative process of distributed scheduling commences. During each iteration, scientists formulate their science *requests* (the tasks they want accomplished) in terms of the modules they would like performed. A particular request may specify all of the parameters of each module, or may specify some parameters and place constraints on other parameters. For example, the particular time at which the module is executed may not matter so long as that time corresponds to night at the receiving earth sta-

---

[1] It should be noted that Voyager schedulers did evaluate the use of AI planning tools for this problem (Vere 1983). These tools were abandoned because of poor search performance.

[2] The specific sets of modules vary with each mission. Hence, the examples of modules and spacecraft constraints provided in this paper should be taken only as a didactic interpretation by the authors. Actual modules may refer to a considerably lower level of detail than our examples suggest.

tion or so long as a particular planet is visible to the instrument at that time. Similarly, the requester may only care that *some* tape recorder is used to store information, in which case she would specify the `record-instrument` module without specifying its tape-recorder parameter.

Given a set of module requests and a history of the outcomes of the previous iterations, the task of the *schedule coordinator* is to produce a schedule for the next iteration which satisfies a set of generally informal criteria. This involves characterizing the various requests in terms of their priority and flexibility, adding highly constrained or constraining tasks which have high priority to the schedule, and asking users to negotiate on requests which may be difficult to achieve as stated. A particularly difficult computational task in this process is that of *schedule refinement*, or converting the partially constrained modules requested by scientists into a more detailed schedule which satisfies the constraints on these requests. This refinement takes place by the following two types of operations:

- **constrain** some parameters in existing modules. These constraints may be local to the module (e.g., constraining the absolute start time of the module) or may be relative to other modules (e.g., constraining one module to start after another has finished).

- **add** new modules to satisfy the preconditions of existing modules. These added modules essentially "glue" different modules together. An example is to add a slewing module between two observation modules. Like the original modules, these added modules need not specify all parameters.

This computational problem is similar in nature to problems treated by standard AI planning and scheduling algorithms, but with some important differences. We believe that this problem is most accurately characterized as *planning in service of scheduling*, in which both planning and scheduling are necessary but in which the major computational difficulty lies in scheduling.

In traditional precondition-achievement planning (Fikes, Hart, & Nilsson 1972), the planner is presented with a set of conditions to achieve, and the task is to find a set of actions to achieve these conditions. In the present context, human scientists do the high-level planning to convert goals into modules, and the task of the program is to refine those modules and add new ones to ensure that the overall schedule is *operational*. In this framework, the precondition-achievement left to the algorithm (to achieve conditions on modules already requested) is a very simple problem, since there are only a small number of ways in which different modules can be combined.

As discussed above, a common form of constraint on a module's parameters will be to constrain relative and absolute times of actions. While this is a key aspect of scheduling problems, this problem also necessitates extensions to traditional scheduling. In particular, most scheduling algorithms (Smith 1994) take as input a set of actions and constraints about the interactions between these actions, and attempt to assign (possibly relative) times to those actions so that the constraints are satisfied. In the current problem, in addition to time of actions, the problem-solver can also alter parameters associated with actions which may take on continuous values and affect state. Another difference is that traditional schedulers do not generate actions, as is necessary in the current problem.

3. Supporting mission operations

In our discussion of mission scheduling so far, we have focused on *computationally* difficult aspects of problem solving (e.g., producing a schedule satisfying a set of constraints). While these aspects must be addressed by a scheduling program, it is equally important to automate or support the whole spectrum of *practical* but important functions performed by real human schedulers. This section outlines the actual process of scheduling as it relates to mission operations more broadly, with emphasis on the types of support needed from planning and scheduling tools, beyond combinatorial problem-solving.

Scheduling in practice involves making complex tradeoffs among often conflicting requirements of different machines and individuals. Human schedulers spend much of their time *explaining* why they made their decisions, and *negotiating* compromises when several tasks cannot all be achieved at once. For example, a scheduler might inform a scientist who requested an observation over a certain time period that her request might cause an instrument to be damaged, or ask her to shorten the requested time period to accommodate an important observation by another scientist.

Once a workable and agreeable initial schedule is produced, it is distributed to the individuals involved in the mission, who then use the schedule as a basis for their own subsequent planning. From this point, the human scheduler's task is to handle new considerations as they arise. The ability to make *incremental and localized modifications* to the existing schedule is crucial. The knowledge that the schedule will be broadly correct enables individuals to use an intermediate schedule as a basis for planning. Non-local changes impact the whole organization, and their cost increases the longer the schedule has been out. Local repair techniques (Minton *et al.* 1992; Zweben *et al.* 1993) may provide a useful basis for low-cost modifications to schedules. Structural domain decomposition provided by localized search techniques (Lansky 1994), may also produce local modifications during re-planning.

This process of iterative scheduling and refinement continues until final commitments are made, which are then *translated* into detailed command sequences. Much effort is spent in this translation process, and in *verifying* that the sequences are safe and satisfy the science schedule.

Even once the sequences are then up-linked to the spacecraft, the job of human schedulers continues in the case of *error recovery* – the case where things did not all go as planned. Here again, it is important to be able to explain what was supposed to be happening and to make hopefully small modifications to the schedule in light of the new circumstances. This functional requirement may overlap that discussed earlier with respect to iterative scheduling, but in this context there is a much higher emphasis on real-time recovery.

In all of these scheduling functions, *robustness* and *flexibility* are desirable properties of a schedule. Inflexible schedules increase the chance that incorporating new considerations will require global changes and increase the likelihood that uncertainty in execution will result in failure (Drummond, Bresina, & Swanson 1994; Muscettola 1994b; Hanks & McDermott 1994).

The process described in the previous section and the desiderata illustrated in this section raise strong requirements on three aspects of any automated solution:

- **language:** A domain-representation language must facilitate the representation of complex, parameterized constraints. These constraints may be based on the function and structure of spacecraft components, which suggests that the language will have to provide sophisticated modeling tools and enable domain models to be developed at multiple levels of abstraction.

- **algorithm:** A planning/scheduling algorithm must be able to operate on constraints and models expressed in such a language. It must also integrate features of planning systems (such as action generation, task reduction, and parameter specification) with features of scheduling algorithms (such as task ordering, metric and resource constraints). In addition, the algorithm must be iterative, interactive and incremental, so that its output at each stage of schedule refinement can be used as a basis for the human actions which are taken concurrently with the scheduling process.

- **external knowledge:** Lastly, both the language and algorithm must enable the user to make use of extensive knowledge stored in other software packages, as it will in practice be impossible to replicate this knowledge. As a simple example, the system must be able to figure out where Pluto will be at any given time in order to interpret the constraint "instrument pointing at Pluto."

We are currently extending the HSTS system to apply it to this domain. The HSTS scheduler embodies a design philosophy which integrates planning and scheduling, and it uses a representation language (HSTS-DDL) which facilitates the development of parameterized domain models, modular structure via an object-oriented representation scheme, and constraint representation at multiple levels of abstraction. In the next section, we discuss this system in more detail.

4. The architecture of HSTS

Figure 1 shows a block diagram for the architecture of HSTS. The two core components are the Domain Description Language (HSTS-DDL) and the Temporal Data Base (HSTS-TDB). HSTS-DDL allows the specification of a model of the domain. In a spacecraft sequencing domain, for example, a model will contain a characterization of all the constraints discussed in Section 1. HSTS-TDB supports the generation of short-term schedules that are consistent with the HSTS-DDL model. By posting assertions and constraints among assertions in the data base, a planner/scheduler sets goals, builds activity networks, commits to the achievement of intermediate states, and synchronizes system components. The tight connection between the entities that can be specified in HSTS-DDL and those that can be represented in HSTS-TDB has two main benefits. First, it

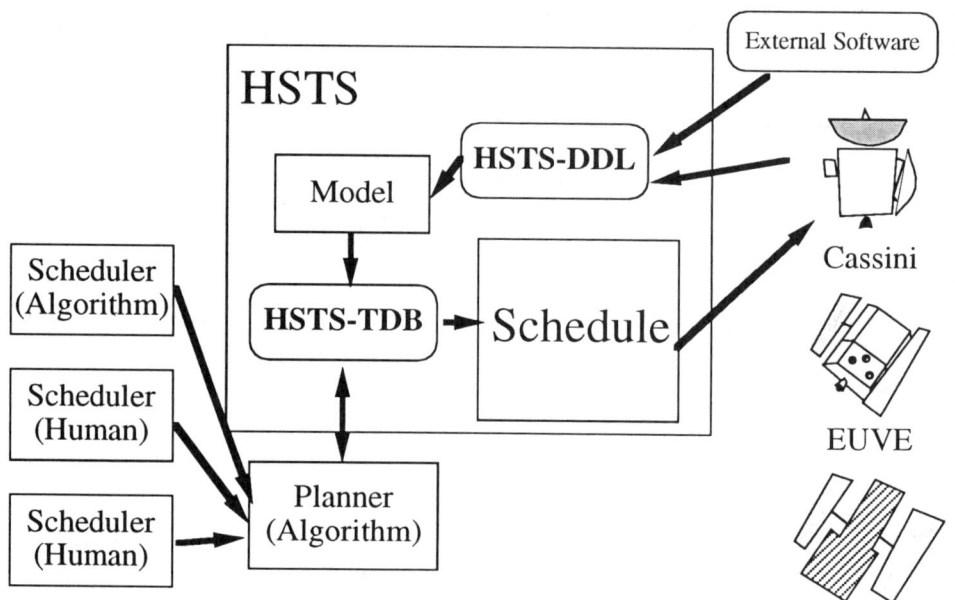

Figure 1. Block diagram architecture of HSTS.

assures compliance to the model. Second, it provides a strong basis for efficient schedule generation by taking advantage of the domain structure during problem solving.

We have already built several scheduling systems for short-term mission scheduling using HSTS. Their general structure involves a phase during which observations are allocated in sequence for execution on the different spacecraft instruments (scheduling) and a second phase during which detailed constraints are enforced and auxiliary activities are generated as needed (planning). To generate a short term schedule a problem solver repeatedly alternates between planning and scheduling. Currently we are using a completely automated planning phase. The scheduling phase is different in different applications and can involve human decision making. Both planning and scheduling phases are customized with heuristics depending on the domain model expressed in HSTS-DDL.

We will now focus on two aspects of HSTS: expressing domain models in HSTS-DDL and integrating HSTS with external software.

4.1. Modeling space mission domains

Accurate models of the spacecraft are essential for the generation and verification of a detailed sequence over an extended time horizon (potentially covering the entire mission duration). A less model-intensive and more reactive planning/execution approach is not feasible since the current generation of spacecraft has a very limited amount of autonomy to react to and recover from anomalous situations. We are building detailed spacecraft domain models for science mission planning by using and extending HSTS-DDL. This domain description language has many features that we believe to be fundamental for the success of an automated planning tool in this domain.

HSTS-DDL follows an object-oriented approach to domain specification. The model of a system is built by assembling models of subsystems. The state of each subsystem can be described by the values assumed by a few state variables over continuous time. This domain structuring technique, also employed by Lansky (1988), enables clearer indexing of the instantaneous state of the system when compared to the relatively unstructured state descriptions used in classical approaches to AI planning.

As an example of state variable declarations, a spacecraft behavior can be described by giving the direction where the spacecraft is pointing and the state of each of its instruments. In HSTS-DDL this can be expressed by the following class declarations.

(Define_Object_Class Spacecraft
  :state_variables
  ((Controllable Pointing_Status)))

(Define_Object_Class Instrument
  :state_variables
  ((Controllable Op_State)))

Instances of these classes can be created to describe different spacecraft (e.g., Cassini) and different instruments on a spacecraft (e.g., VIMS and UVIS on Cassini). An HSTS-DDL model is therefore highly *modular*. Models of increasing complexity can be developed by reusing and assembling smaller models. Likewise, planning/scheduling heuristics can be indexed with respect to the subsystem to which they are relevant. This facilitates both scaling and reuse.

As the previous example shows, state variables (e.g., **Pointing_Status** and **Op_State**) have an associated type (**Controllable**) which determines the way they will be implemented. State variables whose value can be altered by a planner/scheduler are **Controllable**. Their implementation allows the full range of constraint posting and propagation provided by the HSTS temporal data base. Time-lines of fixed events (e.g., when a certain star rises and sets for an earth-orbiting observatory) cannot be manipulated by a planner/scheduler. HSTS provides a reduced (and more efficient) implementation when such a state variable is declared **Uncontrollable**.

Beyond what is illustrated in the above example, strong *data typing* is pervasive to an HSTS-DDL model. Besides enabling increased efficiency in the implementation (an important factor for a large, practical planner/scheduler), data typing facilitates the construction of large models. A standard benefit of typed programming languages is that many errors can be detected at compile time rather than during problem solving. In HSTS-DDL type checking is implemented by a *back-end independent translator* that can translate models into data structures interpretable by a planning/scheduling system. Currently, the HSTS-DDL translator creates Lisp code compatible with the current version of the HSTS temporal data base. A future version will be geared toward a C++ implementation. Conceivably, back-end translations could be tailored for different planning/scheduling systems with semantics comparable to HSTS-DDL.

Each state variable's value typically has a finite duration. As mentioned in Section 2., the scheduling aspect of the problem is central in this domain and cannot be treated as subordinate to the planning aspect, as is often done in the AI planning community. HSTS-DDL explicitly represents *metric time* and *finite capacity* associated with model entities. The value of a state variable evolves over continuous time. State variables generalize scheduling resources by combining capacity with state information.

Another powerful programming technique HSTS-DDL supports is *constraint-based* representation. Models can express interval temporal constraints among state variable values and relations between specific parameters of state variable values, i.e., arguments of the predicates that represent these values. Ideally, constraints should always be expressed *declaratively*, so as not to force a specific model of problem solving (e.g., temporal projection, backward chaining). In practice, it is often difficult to develop a model without already having in mind a problem solving method to use. HSTS-DDL allows the expression of constraints as functions returning specific predicate argument values once values of other temporally related predicates are known. Such functional attachments introduce a definite order in which a planner/scheduler should try to satisfy the model constraints.

To give an example of a complete pattern of temporal constraints, we list below an HSTS-DDL fragment from our current Cassini model (see section 5.2. for a description of the Cassini application). This compatibility rule describes a constraint on the state variable (SSR operating_status), which corresponds to the operating state of Cassini's internal tape recorder. The constraint expresses the conditions under which the tape recorder's state can assume a value (read_out) corresponding to reading out data from the detector (for example, the camera) of some instrument. Important parameters of this read-out state of the tape-recorder are the length of the current exposure, the type of the detector being recorded, and the amount of data already stored on the tape recorder.

The constraint is broken up into three conditions. The argument of the second constraint involving stored_capacity, which corresponds to the final amount of data stored on the tape recorder after this record event, is computed by executing the given Lisp code fragment once the values of the variables are known. This can happen only if temporal projection is used to determine the temporal evolution of the state variable (SSR operating_status), because the Lisp addition function is not invertible.

```
(Define_Compatibility
  ;; Constrained token
  ((SSR operating_status)
   (read_out (?detector ?exp_length ?existing_data)))
 :compatibility_spec
  ;; Constraints on the token
  (AND
    ;; (1) Predecessor value on state variable
    (met_by
      ((SSR operating_status)
       (stored_capacity (?existing_data))))
    ;; (2) Successor value on state variable
    (meets
      ((SSR operating_status)
       (stored_capacity (Lisp (+ ?existing_data
                                 (compute_data_production
                                   ?detector ?exp_length))))))
    ;; (3) Parallel condition on state of another variable
    (contained_by [0 _plus_infinity_] [0 0]
      ((?detector data_status)
       (data_stored (?detector))))))
```

The functional attachment mechanism is aimed at achieving a practical balance between versatility of the model and support to prototype development. A quickly developed but more restricted system can give useful empirical feedback. Such feedback can be incorporated into a more general and flexible model at a later stage.

4.2. Integrating with the User's Environment

External tools and knowledge sources are routinely used by human schedulers when building spacecraft science plans and sequences. If such tools and knowledge are not integrated within an AI planner or scheduler, it is unlikely that the tool will be used in practice. Two important integration issues concern external geometric knowledge and user interfaces. This section discusses these in more detail.

As shown in Section 4.1., HSTS allows the user to make reference to functions not directly interpretable in the HSTS-DDL. This is the natural place where *external software* can be integrated within the model. Our scheduler can make use of extensive external subroutines for different kinds of reasoning. A prime example is JPL's NAIF toolkit, an extensive FORTRAN library of geometric subroutines used for planetary calculations. The routines enable scientists to determine, for example, how long light will take to travel from a source to an instrument, or whether a given planet will be visible to the spacecraft at a specific point in its mission. To facilitate the use of these routines, we use AMPHION (Lowry *et al.* 1994), an automatic programming tool which compiles graphical, high-level, declarative specifications into a sequence of NAIF subroutine calls.

A second integration issue involves graphical user interfaces. We are exploring ways in which HSTS could work with a variety of available user interfaces, rather than requiring a new interface to be developed for each scheduler or mission. This leads to the need to consider the functionalities common to the planning or scheduling task itself, and has prompted us to work toward an *interlingua* through which schedulers and generic scheduling interfaces can communicate without knowing the details of the other program.

HSTS currently uses DTS, the customizable scheduling interface being developed by HRI (Hansson & Mayer 1994). By "interlingua," we mean that HSTS and DTS communicate at the abstract level of scheduling data, not at the detailed level of how the data should be displayed to, or manipulated by the user. When HSTS updates the start-time of an exposure, it sends DTS a simple message to that effect, and DTS decides how this should affect the display.

The overall goal of this integration effort is to provide a scheduling tool which is embedded in the user's environment and which makes use of the knowledge already there, provides a common look and feel across tasks, and is customizable to the needs of the user. This type of integration will be necessary for any planning or scheduling tool to be used in the real world.

5. Applications

HSTS is currently being used to support a number of space scheduling applications. The main applications at present are the Extreme Ultraviolet Explorer (EUVE) and Cassini. We are also gearing up to use HSTS to support NASA's New Millennium spacecraft project, which will apply advanced spacecraft technology to enable new mission scenarios with radically decreased operations costs.

5.1. EUVE

The Extreme Ultraviolet Explorer (EUVE) (Bowyer 1990; 1994) is an earth-orbiting telescope which makes observations in the wavelength range of 70 to 760 angstroms. The EUVE is operated by the NASA Goddard Space Flight Center and the Center for EUV Astrophysics (CEA) at the University of California, Berkeley. In the upcoming EUVE Extended Mission, CEA will take over all possible operations responsibility for the spacecraft, without corresponding budget increases. Because automation is the only means toward this ambitious goal, EUVE has become a testbed for automated low-cost operations in future NASA missions.

Automated science scheduling is one important step in automating EUVE operations. We have installed a manual science scheduler at CEA, which assists their experts in generating science plans (observing schedules at the level of pointings, rolls, instrument configuration and exposure times). By the end of 1995, the science scheduling process will be fully automated.

The constraints in this problem are determined by the positions of observational targets, the position and attitude of the spacecraft platform, and the positions of obstacles such as planets, the sun and atmospheric anomalies. A number of CEA preprocessors check whether these constraints are violated by combinations of pointings, rolls and desired exposure times. More complex observations, including dithered targets and moving targets, require quite sophisticated constraint-checking. For moving targets, the constraint-checkers can require over a CPU day to verify that an exposure will not violate spacecraft constraints. These detailed checks by the science planner are vital because if a constraint-violation were to be detected by the spacecraft instead, it would jeopardize spacecraft health *and* cause the spacecraft to enter a "safe-hold" position, causing lost science time and increased operations costs.

Many scheduling systems take an extremely narrow view of the scheduling process. They schedule a fixed set of observations, under prespecified constraints. But the EUVE problem is very dynamic: it involves variable slew times, multiple exposures, pointings that match an ephemeris, etc. The problem also requires quite idiosyncratic constraint-checking by a number of external programs, as well as integration with a number of proposal databases, star catalogs, etc. By one recent count, CEA uses 14 separate programs during its scheduling process. Inefficiencies result because these 14 programs are run in a manually-controlled "pipeline." By integrating these programs within HSTS, we are not only automating the process, but in fact, reducing the computational requirements, as these auxiliary programs are invoked *as needed* to provide information or check constraints.

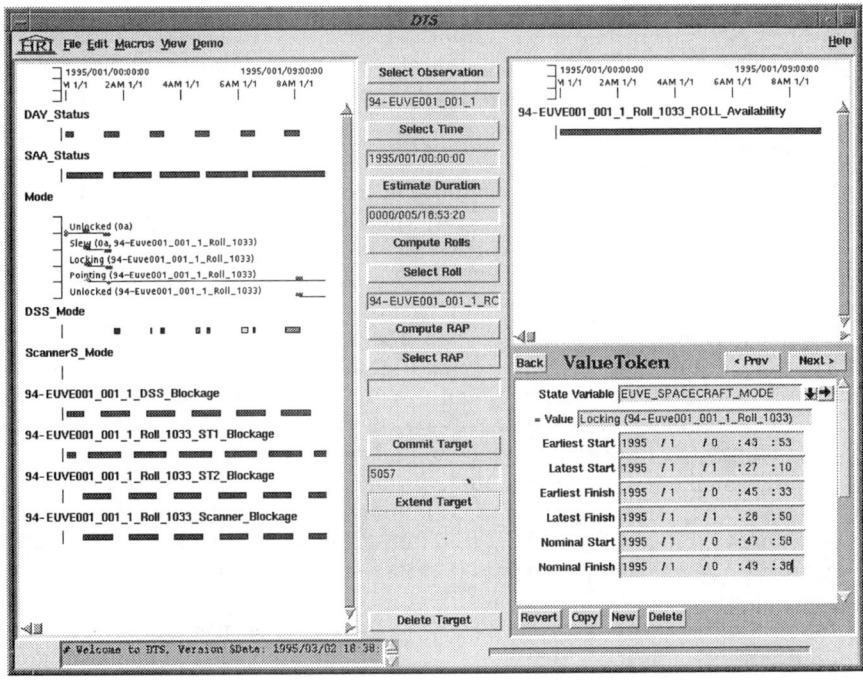

Figure 2.   The EUVE scheduler.

Finally, using the DTS interface, we are attempting to capture the scheduling process that has been developed at CEA. Figure 2 shows the interface developed for EUVE. This interface was customized for use by CEA within one man-month of programmer effort.

The screen is dominated by two DTS "dynagraphs" (dynamic graphs) displaying timelines. The left dynagraph shows details of the current schedule. Reading from the top of the screen, the interface displays day/night status, South Atlantic Anomaly (SAA) status, spacecraft mode (e.g., pointing, slewing, etc.), exposures, and then various star-tracker and instrument blockages. The right dynagraph displays information on whether the possible spacecraft roll angles violate constraints. The lower-right portion of the screen is a form displaying details on objects selected by the user using the mouse: for example, when an exposure is selected, this form displays start-times and end-times to the second.

In the center of the screen is a vertical column of buttons, which embody the typical scheduling process. From the top, the buttons permit the user to (1) select a target to schedule, (2) choose an earliest start-time or a gap in the schedule, (3) invoke an external program to estimate elapsed time to perform the observation, (4) compute constraint-checks on all possible roll angles, (5) choose a specific roll angle, (6-7) compute and select possible targets to select in another spacecraft instrument, and finally, (8) commit the observation of this target to the schedule. This "fill-in-the-blanks" representation of the science scheduling process has greatly simplified the interface.

5.2. Cassini

Cassini is a large spacecraft with destination Saturn and Titan. The project is undergoing joint development by NASA, the European Space Agency and the Italian Space Agency. The U.S. portion of the mission is managed for NASA by the Jet Propulsion Laboratory. The date of launch is October 1997; after 7 years of cruise, Cassini will arrive at Saturn in June 2004 and will orbit the planet some five dozen times over a 4 year mission. During this time, Cassini will perform experiments toward several science objectives including the exploration of Saturn's and Titan's atmosphere and the study of Saturn rings and moons.

While the number of instruments carried by the Cassini spacecraft (12) is comparable to that of Voyager, the mission has much higher complexity. Cassini is expected to make at least 50,000 observations during its 4 year mission. While the amount of science to be performed will increase by almost two orders of magnitude, budget limitations are likely to reduce the number of support schedulers by an order of magnitude, leaving scientists responsible for most of their own sequencing.

In addition to higher complexity due to increased science and reduced personnel, Cassini is also more complex than previous missions because of the on-going re-planning necessitated by Cassini's continuous observations. Experience has shown that periods of intense observation give rise to the highest workload and stress levels for scientists and schedulers. This is because new information becomes available which was not present at the time the initial schedule was constructed. In such cases, re-planning may be necessary either to avert a potential failure of existing goals (as when a spacecraft drifts too much from its original course) or to take advantage of unanticipated science opportunities (such as the discovery of rings around Neptune, which happened after the main scheduling was done for the Voyager Neptune encounter). Whereas previous missions were characterized by short periods of intense observations and, hence, limited periods of re-planning (when the spacecraft was near a planet), intense observations for Cassini will be the characteristic mode of operation during its four-year mission. Throughout all of this period, scientists will be subject to continuous stress caused by the need for re-planning.

In summary, while many missions had large staffs, smaller science objectives, and shorter periods of high-pressure re-planning, the Cassini mission will have a greatly reduced staff, greatly increased science objectives, and a four-year sustained period of high-pressure re-planning. While scheduling each Voyager encounter could be seen as four years of relative calm followed by three weeks of panic, without advances in mission operations software, scheduling Cassini is likely to be four years of total panic.

We are currently collaborating with the Cassini mission and more specifically we are supporting the activities of the Cassini rings and dust science working group. Although the construction of the spacecraft has not been completed and several details of its behavior are still unknown, scientists are already evaluating orbital geometries with respect to their potential scientific output. An essential tool of this evaluation is the generation of representative schedules of experiments and observations. To this end, we have used HSTS to build a scheduler for the Cassini mission (Figure 3). The scheduler has been used to generate a representative schedule of observations of the Saturn rings during the 14th orbit of Cassini around Saturn. The current HSTS-DDL model of the Cassini domain keeps track of the visibility of the Deep Space Network (DSN) stations (Goldstone, Canberra and Madrid) and the occurrence of other significant mission events, such as star occultations and flybys of Titan and other satellites. As with EUVE, we model the observing state of all instruments and the slewing and pointing state of the spacecraft. Additionally we model the storage of scientific data in the on-board solid state recorders and the down-link of data to Earth through DSN stations.

5.3. New Millennium

NASA has recently announced the New Millennium program. The program will take the form of a series of aggressive technology-demonstration missions. The goal of the program is to open a new mode of space exploration. This new mode will consist of "faster, smaller, cheaper" spacecraft (as described by NASA Administrator Dan Goldin), with the goal of eventually establishing a "virtual presence" in space.

With an emphasis on low-cost operations to support a fleet of spacecraft, New Millennium is encouraging the application of advanced automation techniques to support ground operations and on-board autonomy.

In terms of science scheduling, one proposal is to move much of the functionality of scheduling and sequencing on-board the spacecraft. In this model, the scientists would communicate high-level science goals directly to the spacecraft. The spacecraft would then perform its own science planning and scheduling, translate those schedules into sequences, verify that they will not damage the

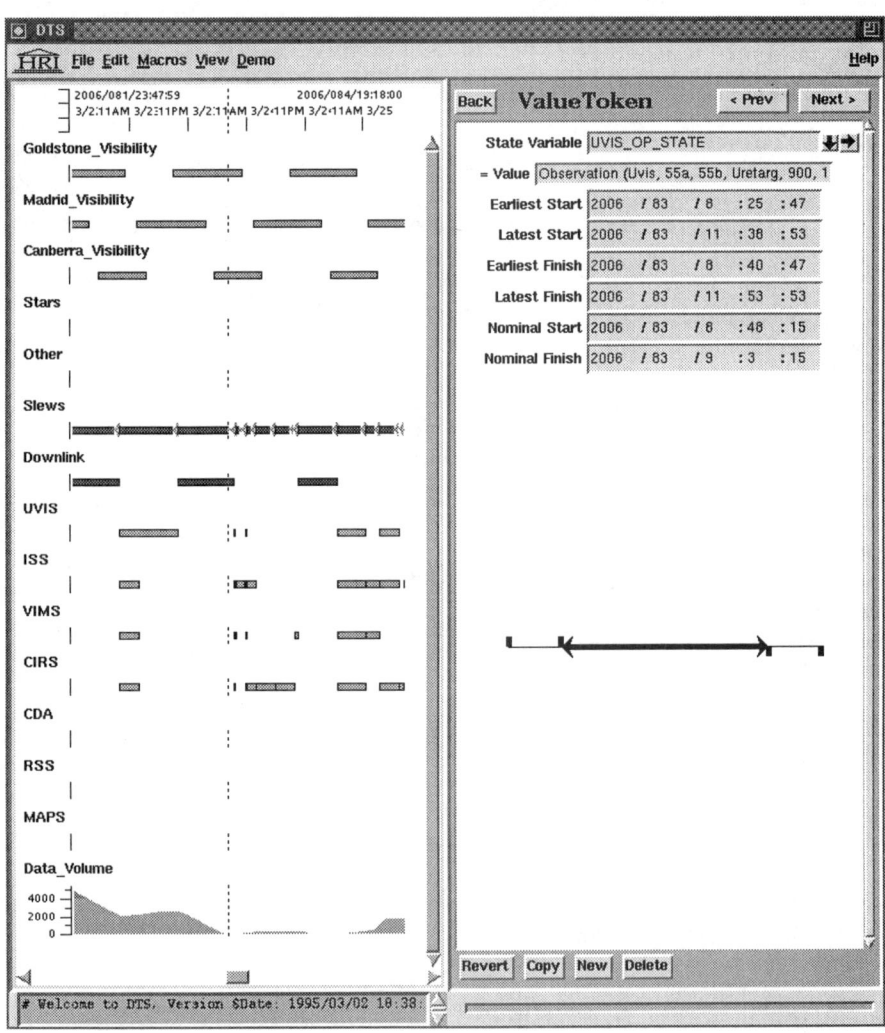

Figure 3. The Cassini scheduler.

spacecraft, and ultimately execute them without routine human intervention. In the case of error recovery, the spacecraft would have to understand the impact of the error on its previously planned sequence and then reschedule in light of the new information.

We are currently investigating ways to extend HSTS to meet the challenges of increased autonomy and reliability within a preliminary demonstration of an autonomous mission concept. Any automated solution will require strong modeling capabilities, powerful algorithms for constraint processing and search, and facilities for integration with a multitude of external software routines. We are confident that the development of these components of HSTS in support of existing missions like EUVE and Cassini will enable increasingly automated scheduling for New Millennium.

6. Conclusion

This paper has described our experience to date in developing a planning and scheduling framework, HSTS, to support scientific space missions. Space mission scheduling is a good example of a practical problem with important repercussions on the cost and, therefore, the feasibility of future space exploration.

The automation of space mission scheduling poses several challenges to computer science. Key among these are supporting the entire process of scheduling rather than just a portion of it, and developing powerful modeling languages that can deal with large models and diverse sources of information. We have addressed these challenges in applying HSTS to EUVE and Cassini mission scheduling.

Acknowledgements

Thanks to Roger Malina, Eric Olson and Gary Wong for their role in the EUVE application and to Jeff Cuzzi and Ken Bollinger for serving as our domain experts in the Cassini application. Thanks to Mike Lowry, Andrew Philpot and Tom Pressburger for assistance with using AMPHION. We are grateful to John Bresina and Lise Getoor for useful comments.

References

AAAI. 1994. *Proceedings of the Twelfth National Conference on Artificial Intelligence*, Seattle, WA: AAAI/MIT Press.

Atkinson, D., ed. 1994. *Proceedings of the Third International Symposium on Artificial Intelligence, Robotics, and Automation for Space (i-SAIRAS)*.

Bowyer, S. 1990. The extreme ultraviolet explorer mission. In Kondo, T., ed., *Observatories in Earth Orbit and Beyond*. Kluwer Academic Publishers. 153–169.

Bowyer, S. 1994. Extreme ultraviolet astronomy. *Scientific American* 271(2):32 ff.

Bresina, J. L. 1994. Telescope loading: A problem reduction approach. In Atkinson (1994).

Drummond, M.; Bresina, J.; and Swanson, K. 1994. Just-in-case scheduling. In *Proceedings of the Twelfth National Conference on Artificial Intelligence* (1994), 1098–1104.

Fikes, R.; Hart, P.; and Nilsson, N. 1972. Learning and executing generalized robot plans. *Artificial Intelligence* 3(4):251–288.

Fox, M., and Zweben, M., eds. 1994. *Intelligent Scheduling*. Morgan Kaufmann.

Hanks, S., and McDermott, D. 1994. Modeling a dynamic and uncertain world i: Symbolic and probabilistic reasoning about change. *Artificial Intelligence* 66(1):1–55.

Hansson, O., and Mayer, A. 1994. DTS: Building custom, intelligent schedulers. In Atkinson (1994).

Johnston, M. D., and Miller, G. E. 1994. SPIKE: Intelligent scheduling for hubble space telescope observation. In Fox and Zweben (1994).

Lansky, A. 1988. Localized event-based reasoning for multiagent domains. *Computational Intelligence Journal* 4(4). Special Issue on Planning.

Lansky, A. 1994. Localized planning with diverse plan construction methods. Technical Report FIA-94-05, NASA Ames Research Center.

Lowry, M.; Philpot, A.; Pressburger, T.; and Underwood, I. 1994. Amphion: Automatic programming for scientific subroutine libraries. In *International Symposium on Methodologies for Intelligent Systems*.

Minton, S.; Johnston, M. D.; Philips, A. B.; and Laird, P. 1992. Minimizing conflicts: a heuristic repair method for constraint satisfaction and scheduling problems. *Artificial Intelligence* 58:161–205.

Muscettola, N. 1994a. HSTS: Integrating planning and scheduling. In Fox and Zweben (1994).

Muscettola, N. 1994b. On the utility of bottleneck reasoning for scheduling. In *Proceedings of the Twelfth National Conference on Artificial Intelligence* (1994), 1105–1110.

Smith, S. F. 1994. OPIS: A methodology and architecture for reactive scheduling. In Fox and Zweben (1994).

Vere, S. 1983. Planning in time: Windows and durations for activities and goals. *IEEE Transactions on Pattern Analysis and Machine Intelligence* 5:246–267.

Zweben, M.; Davis, E.; Daun, B.; and Deale, M. 1993. Scheduling and rescheduling with iterative repair. *IEEE Transactions on Systems, Man, and Cybernetics* 23(6):1588–1596.

Scheduling the IUE Satellite with Constraint Logic

Bruce McCollum

Science Programs, Computer Sciences Corporation, Code 684.9, NASA Goddard Space Flight Center, Greenbelt, MD 20771

Mark Graves

Department of Cell Biology, Baylor College of Medicine, One Baylor Plaza, Houston, TX 77030

Abstract. The International Ultraviolet Explorer (IUE) satellite observatory carries out several thousand different science observations each year. These observations must be scheduled under a variety of hardware-dependent and science-dependent constraints. A typical observation is affected by some but not all possible constraints. Some constraints also change over time, sometimes unpredictably. We describe these constraints and strategies used to simplify the scheduling problem. We also report on constraint-logic based artificial intelligence software which we have developed to serve as a high-level aid to scheduling the IUE. The algorithm is sufficiently simple that the scheduling can be done by one person with a standard workstation. This software, in modified form, is potentially applicable to scheduling other astronomy satellites. Our implementation permits a large degree of scheduling flexibility, including the ability to change the schedule on short notice. Also, the use of constraint logic permits adding different kinds of constraints without also having to implement new methods for satisfying them, so that it is relatively easy to modify the software for changing needs.

1. Introduction

The IUE (International Ultraviolet Explorer) satellite is an orbiting observatory which has been operating continuously in geosynchronous Earth orbit since 1978, run jointly by NASA and the European Space Agency. Each year over a hundred different science programs have been carried out. Scheduling observations on the instrument must take into account many different constraints, some due to hardware and satellite operations limitations, others due to the science requirements of specific programs. Until recently, IUE science has been scheduled with minimal software support. We are developing a new artificial intelligence package which serves as a sophisticated aid to scheduling this satellite, and which has potential uses in scheduling other astronomy satellites.

2. Description of the Scheduling Problem

Over a hundred different NASA science programs are carried out every year by the IUE, consisting of over 4000 distinct observations. Scheduling them is limited by two broad categories of constraints. One category consists of constraints due to aspects of the hardware and satellite operations. The other consists of constraints due to the science requirements of a particular science program to be carried out.

Hardware and operations constraints are independent of the science programs and are always in effect, although the same constraint may produce different limitations on different programs. These constraints include:

1. The pointing angle of the telescope relative to the direction of the Sun must always remain within a range of values (nominally 28 degrees to 108 degrees). This is in order to keep the solar power cells nearly perpendicular to the Sun so that enough power can be generated. This constraint has the effect that most objects in the sky can be observed during only part of the year. The specific days during which a target is observable depends on the target's location in the sky.

2. There are four different commonly used modes of taking data. This is due to the existence of two detectors which cover different wavelength regions, each capable of two different spectral resolutions. Only one mode at a time can be used.

3. For 8 hours of every day, there is high background radiation due to the path of the satellite's orbit. During this time, observations requiring long continuous exposures of the same target cannot be done, since the background radiation shows up as noise in the data. This limits the time of day, but not the time of year, when some observations can be scheduled.

4. From November through February, the telesope cannot be kept pointed for more than about 4 hours within a certain range of angles relative to the Sun. This is due to gradual overheating of the on-board computer due to solar heating. The range of acceptable angles varies from month to month.

5. The telescope cannot be rolled (by any significant amount) about the pointing axis. This affects some programs because the main aperture of the telescope is elliptical. For some science programs, the orientation of the aperture relative to the target is important. The orientation of the aperture against the sky does rotate through 180 degrees every 6 months as the Earth goes around the Sun. For this reason, science programs which need a certain aperture orientation are constrained to certain times of the year, depending on the specific orientation required.

6. NASA programs can be scheduled only during 16 hours of each day. During the other 8 hours, the satellite is operated by ESA to carry out European science programs.

We note that since the IUE is in a high Earth orbit, the Earth rarely blocks a target, and such blocking is of short duration when it occurs. The problem of the Earth blocking a target thus rarely affects IUE science scheduling. For satellites in low Earth orbit, where the Earth occupies a much larger fraction of the apparent sky, this is a much more frequent and serious constraint.

Scheduling constraints which depend on the particular program being carried out can include any or several of the following.

- Some observations require low background radiation, which limits the time of day during which they can occur.

- Some observations require certain exact times of the year

- Some observations must take place on certain days of the year, but with a tolerance of a few days

- Some science programs require observations spaced at certain intervals throughout part of the year (with or without a tolerance of a few days)

- Some science programs require a single continuous series of observations for a few days or more

- Some observations are part of joint NASA-ESA science programs, which must be scheduled contiguously (most NASA programs can be scheduled independently of the ESA schedule)

- Some observations must be scheduled simultaneously with observations on another instrument (e.g. a ground-based telescope or another satellite)

- Some observations must occur within some given time interval of one another, but the time of year is not constrained by the science

- Some observations must be scheduled on short notice at unpredictable times ("target of opportunity" programs to observe novas, supernovas, X-ray transients, etc.)

We note that observations (an observation being defined as a period of time spent observing one target with one instrument configuration) vary in length from one second to 16 hours. The average length of an observation is in the neighborhood of 20 to 30 minutes. In between each observation, certain "housekeeping" tasks must be performed for operations reasons, which can take from a few minutes to over 40 minutes per observation, depending on the nature of the observation. E.g., pointing the telescope to a new target and verifying that the target is centered in the aperture can take a few minutes to over an hour, depending on the location and brightness of the target. Since data cannot be acquired during these operations tasks, additional "overhead" time must be taken into account in filling the schedule.

3. Techniques and Solutions: Coarse-Graining and Constraint Logic

There has been minimal software support for scheduling the IUE before the implementation of our AI scheduling routine. Since the beginning of IUE science operations, the problem has been simplified by dividing satellite time into large (8-hour) time blocks. Each time block is filled by one or more observations belonging to the same science program. Planning the exact sequence of observations within a time block is done by the guest investigator whose science is being carried out, with the help of IUE staff members. This is called a "coarse-grained" schedule, since the schedule is composed of "coarse" time blocks treated as "black boxes" in which details on a finer scale are ignored for scheduling purposes. Thus instead of having to schedule 4000 time blocks per year, the scheduler must arrange 730 time blocks, in keeping with the various constraints listed above.

One advantage of "coarse-graining" is that the scheduler does not have to determine a schedule which is the most efficient in terms of combining operations "housekeeping" and data acquisition. Each guest investigator can plan the sequence of events within his/her time block, selecting targets and adjusting exposure times in order to maximize observing efficiency. This considerably simplifies the scheduling problem. IUE guest observers supervise the acquisition of their data in real time, either in person or from remote sites. This permits rapid changes to be made within a single time block (e.g., changing the planned exposure time of a target) without affecting other time blocks assigned to other science projects.

Another advantage is that coarse-graining increases the observatory's ability to change the schedule on short notice with minimal impact on the overall schedule, because the time blocks are usually mutually independent. If the annual schedule consisted of an uncompartmentalized sequence of observations, then changing a single observation would ripple through the rest of the schedule. Changing the schedule on short notice is sometimes necessary in any satellite observatory because of "targets of opportunity" or hardware failures which may abruptly alter operations constraints.

We have developed artificial intelligence software which incorporates the coarse-grained approach and IUE constraints (Graves and McCollum, 1994). The software combines operations and science-related constraints using constraint logic embedded in the language LIFE (Ait-Kaci, 1991), which is an especially useful language for expressing constraints in a declarative manner. The operations and science-related constraints are used to place constraints on the possible days of the year in which each observation can be scheduled. The constrained observations are then combined into constraints on what days a particular science program (set of observations) can be scheduled for its time blocks. The algorithm then begins sorting through possible combinations of time blocks, i.e. schedules.

In our algorithm, constraints on the observation date restrict the schedules which are considered, which reduces the search space. After the reduced search space is determined, a simple logic program with backtracking is used to create a list of possible schedules and present them to a human scheduler. The search is made faster by considering observations in order of the severity of constraints on possible observation days, with the most time-restricted observations placed in the schedule first. These priorities constrain the search space still more, and so on until a complete schedule or set of alternative schedules is built.

Constraint logic programming is a merger of two computational approaches to problem solving: constraint satisfaction and logic programming. Constraint satisfaction views a problem as a domain and a set of constraints. For example, a constraint satisfaction problem in the domain of integers is to find two integers which are less than 7 and whose sum is greater than 10. Constraints are specified as a set of variables and expressions they must satisfy. Constraint satisfaction is the process of finding a solution within a domain given a set of constraints that the solution must satisfy.

Logic programming views a problem as a program specified by a logic program. In logic programming, a program is defined to be a logic formula. A logic formula is a mathematical expression composed of variables and logical operators such as "and", "or", and "not". The formula is evaluated based on the program inputs to generate the output. Often the input is underspecified and many outputs are possible. When this is the case, each possible output is presented. One advantage of logic programming is that the variables in the formula may be used as either input or output, and thus a logic program can be "run in reverse". A trivial logic program is $Sum(X, Y, S) \equiv S = X + Y$ which can be used for addition, e.g. $Sum(3, 4, S)$ or subtraction, e.g. $Sum(3, Y, 7)$. The real gain of logic programming occurs when they are "run sideways", for example the expression $Sum(X, Y, 10)$ can be used to list all pairs of integers whose sum equals 10.

Constraint logic programming provides a mechanism for solving constraints within the framework of a high-level programming language. Within constraint satisfaction, there are typically a limited collection of domains and operators which can be used. Within constraint logic programming, constraints can be placed on any variables which occur in a logic formula.

We have chosen to use constraint logic programming because it is easier to develop more complex systems of constraints. Implementing a scheduling program consists of defining the domain, specifying the constraints on the domain, and testing possible solutions to see if they meet the constraints. Each one of these steps is fairly straightforward and development time can be spent on insuring that the domain is represented appropriately and that the constraints are complete.

Our algorithm works by modeling the domain of investigator programs, observations, instrument exposures, and targets. The two main constraints on when an observation is scheduled are the request of the investigator and the angle the target has with relation to the sun. Constraints are placed on the observations such that the observations can only be scheduled for days which meet the investigator's request and the target's sun angle. A program is chosen and all of its constraints are checked against the possible days. It is tentatively scheduled for one of the acceptable days, and the process repeats. If a conflict arises, one of the earlier decisions is changed and scheduling continues.

Other satellite scheduling programs exist which use constraints, but none make use of the coarse-grained approach. Some, such as SPIKE (Johnston, 1990) use constraint satisfaction to create "fine-grained" minute-by-minute schedules. However, constraint-logic programming has an advantage over constraint-satisfaction programming in that constraint logic programming deals with constraints in the declarative framework of a logic programming language. Thus, our implementation can be improved by adding different kinds of constraints without having to also implement new methods for satisfying them.

In summary, we have developed a scheduling algorithm which incorporates the traditional IUE "coarse-grained" approach, which can be relatively easily modified to handle new constraints or changes in old ones, and which retains the IUE's ability to reschedule science observations rapidly with minimal impact on the overall schedule. This algorithm is sufficiently flexible that, with some modifications, it could be used to generate schedules for other types of satellites or for ground-based telescopes.

References

Ait-Kaci, Hassan 1991, "An Overview of LIFE", in *Next generation information system technology: Proceedings of the First International East/West Data Base Workshop*, Vol. 504 of *Lecture Notes in Computer Science*, eds. J.W. Schmidt and A.A. Stogny, Springer-Verlag, p. 42

Johnston, Mark D. 1990, "SPIKE: AI Scheduling for NASA's Hubble Space Telescope", in *Proc. 6th Conference on AI Applications*, IEEE, p. 184

Graves, M. and McCollum. B. 1994, "Coarse-Grained Scheduling for Astronomy Satellites", in *Knowledge-Based Artificial Intelligence Systems in Aerospace and Industry*, SPIE Technical Conference No. 2244, eds. W. Buntine and D.H. Fisher, p. 1

PLANNING AND SCHEDULING FOR THE HUBBLE SPACE TELESCOPE

GLENN E. MILLER
Space Telescope Science Institute, 3700 San Martin Dr., Baltimore, MD 21218

ABSTRACT

The Space Telescope Science Institute (STScI) conducts the scientific operations of NASA's Hubble Space Telescope (HST). This paper describes several innovative systems which have been developed to support science operations, including proposal selection, planning and scheduling. The application of these techniques to other observatories and spacecraft is discussed as well.

INTRODUCTION

Figure I illustrates a general model for observatory planning and scheduling. The start of the process is an idea or question which can be addressed by some observations. This is embodied in a scientific proposal which describes the problem, its significance, the observations and how they will be used to solve the problem. Telescope time is a scarce resource and there is usually a competition for time. The proposal is submitted to a review process (which usually includes a peer review). If the proposal is not accepted, the observer must try again, either at a later time or at another observatory. The proposal may be accepted as-is or may be limited in some way (e.g. a subset of the proposed targets or observing time). Planning is the process of defining the how the observations are made while scheduling is the process of deciding when to make them. Planning includes factors such as the instrument, filter, exposure time, ordering of exposures, etc. Planning and scheduling are each performed at different levels of detail, initially from a high level to increasingingly finer detail. The arrow to the left of planning and scheduling in Figure I emphasizes that this is not a linear process. For example, the time of year an observation is scheduled can affect the target acquisition or observing strategy, or the results of an observation can have an effect on subsequent observations. Changes to the observing program must be accomodated by the planning and scheduling process.

This general model of observatory planning is expressed in many different ways at different observatories, ranging from small, ground-based observatories where the observer may be in direct control of essentially all steps, to large, complex observatories where many people support the process. To conduct HST operations, the STScI has developed a number of tools to support all stages of the observing process and has modified these tools on the basis of operational

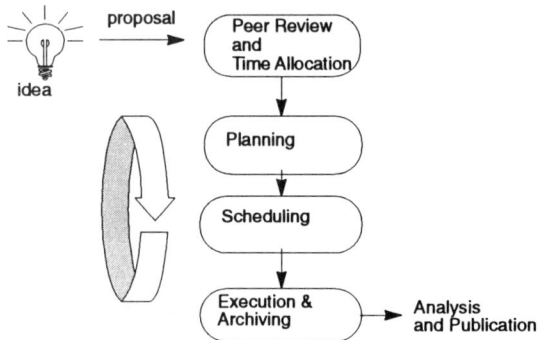

FIGURE I   Schematic of steps in observatory planning and scheduling.

experience.

PROPOSAL SELECTION

From the observer's point of view there are two phases in the HST proposal process. Phase I is the competitive phase culminating in proposal selection while Phase II deals with planning, scheduling and execution. Since HST time is heavily oversubscribed (about three times as many proposals are submitted than are accepted and about eight times more time is requested than available), the Phase I process was designed to place the least possible burden on the proposer and reviewers while providing a thorough and fair review.

For the most recent HST observing cycle (Cycle 5), proposal submission is based on a LaTeX template (Asson 1995). To receive the template, an interested scientist sends electronic mail to newprop@stsci.edu with a subject of "request templates". A computer process monitoring mail automatically returns the template and instructions via email. (Any message which is not recognized as a request for templates nor as a proposal causes a short help message to be returned to the sender.) Other observatories (e.g. Kitt Peak, ESO) have or are considering similar modes of proposal submission.

Instructions are embedded in the template as LaTeX comments. The proposer edits the template to add information such as the name of the investigators, scientific justification, target and observation information, etc. The proposer submits two paper copies of the proposal to the STScI and also electronically mails the LaTeX source file to newprop@stsci.edu. (Most proposers generated the paper copy by running the LaTeX processor on the template but any word processing system can be used as long as the output follows the STScI format.) At the STScI, key information is extracted from the LaTeX source file (proposer's name and institution, target list, instrument usage, etc.) by the mail monitoring process and is entered into a proposal tracking database which is used by the referees and STScI staff.

For Cycle 5, the STScI experimented with fully electronic submission where a Postscript copy of the proposal was submitted in lieu of the paper copy. The results of this were very positive and there was little difficulty in successfully printing the Postscript files at the STScI.

In addition to specifying the targets and observations, the proposer must estimate the number of HST orbits required to execute the observations. In earlier cycles, the proposer estimated this resource usage with the aid of a computer program provided by the STScI. This program read the proposal file to determine the targets, observations, instrument usage, etc. and applied a simple formula for the correlation between these parameters and the amount of spacecraft time used. Proposers would often have to apply correction factors to the output to account for special situations. In Cycle 5 a different approach was used. Using a simple, graphical layout of an HST orbital period and a table of relevant overhead times (guide star acquisition, re-acquisition, instrumental times), the proposer lays out the observations on the graph to determine the total number of orbits required. This simple approach gives the proposer greater insight into using the HST and provides much more accurate resource estimates than the previous procedure.

PLANNING

Once the observing program is approved, each proposer must prepare a more detailed definition of the observations to be made. This is the "Phase II" proposal process and is basically a planning process (see Figure II). Since there is little real-time control of the HST and observations must be fully planned about one month prior to execution, there are many parallels between planning and scheduling robotic telescopes (Drummond, Swanson and Bresina 1994, Genet et al. 1991) and the HST.

The Phase II proposal contains information on the astronomical objects, individual exposures, instrument parameters and the relationships (i.e. constraints) among exposures. This information is contained in an ASCII file which is in a simple keyword-value format. Proposals are submitted electronically using the STScI's Remote Proposal Submission System (RPSS). This software *validates* the contents of the proposal file and can detect a wide range of problems including typographical errors (e.g. misspelled filter name), values out of range (e.g. target declination exceeding 90°), and missing or inconsistent information (e.g. an exposure referencing an undefined target). A dedicated RPSS computer is available to the astronomical community via the Internet. The validation software has also been distributed to approximately 100 astronomical institutions around the world and is run by most proposers at their home institutions before they send the proposal to the STScI.

RPSS was the first system of its kind for a major scientific installation - it has been in use since early 1986. The ability to locally validate and electronically submit a proposal is an extremely valuable tool, both for proposers and the STScI. Proposers can detect and correct a large class of errors and are assured

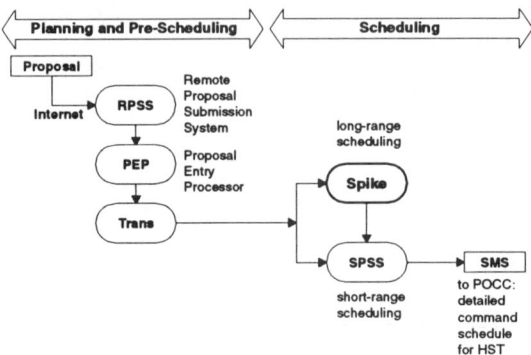

FIGURE II   Processing flow for HST planning and scheduling (see text).

that typographical errors and not introduced by data entry personnel. A further discussion of the proposal handling process can be found in Jackson et al. (1988) and Adorf (1990).

The Phase II proposal describes the observations at a high level. The next step is to describe the observations in sufficient detail to enable their execution on the spacecraft. This step is called *transformation* (Gerb 1991) and involves the determination of instrumental overhead times, grouping observations to minimize overhead operations, choosing specific implementation scenarios and supplying values for instrumental settings that were defaulted by the proposer. Transformation also detects certain errors in the proposal including conflicting timing requirements among exposures, loops in precedence constraints, and inconsistencies in instrumental parameter settings. Transformation makes use of the Spike temporal constraint mechanism (see next Section) to collect and propagate temporal constraints.

SCHEDULING

The *Spike* system (Johnston and Miller 1994) is a general framework for scheduling which was developed by the STScI for HST planning and scheduling. Spike incorporates novel approaches to both the quantitative representation and propagation of scheduling constraints and to efficent construction of nearly optimal schedules.

The HST scheduling problem ranks among the largest and most complex scheduling problems faced on a continuing basis: some 10,000 to 30,000 observations are scheduled per year and each is subject to a large number of operational and scientific constraints (Figure III, see Miller 1989). In the proposal file, observers can specify a variety of constraints on exposures in order to express

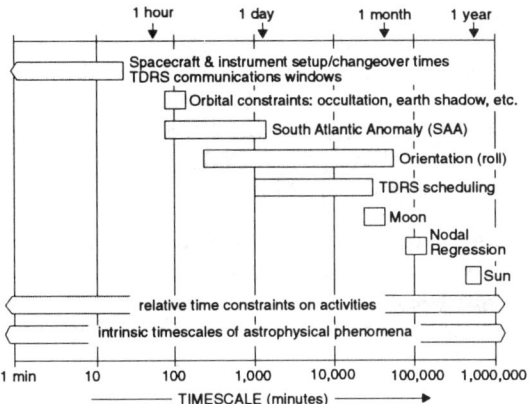

FIGURE III The range of timescales for HST scheduling constraints, covering more than six orders of magnitude

scientific goals: these include relative timing requirements such as precedence, minimum and maximum time separations, ordering, interruptibility and repetition. For example, some observations must be executed within a certain absolute time interval, an observer may constrain the orientation of an instruments aperture relative to the target or require an observation to be made while the HST is in the Earths shadow.

The HST spacecraft and its associated ground system components introduce a number of scheduling constraints as well. The observatory is in a low earth orbit (590 km) with the result that the Earth typically blocks the line of sight to a target for slightly less than half of each 96 minute orbit. Targets within a few degrees of the orbital poles are not occulted by the Earth and are suitable for uninterrupted observations. Due to precession of the orbit, the orbital poles move around the sky with a 56 day period, with the result that a target is available for extended observation for no more than about 3 days in each precessional period. When passing over South Atlantic Ocean, the HST encounters a portion of the Earths radiation belt (called the South Atlantic Anomaly, or SAA) during which instrument operations must be suspended. Sources of bright light such as the Sun, Moon and illuminated Earth must be avoided. Thermal and power restrictions limit how the telescope can be oriented, in order to keep adequate sunlight on the solar panels and off the surfaces which radiate heat.

The primary resource constraint for HST scheduling is the amount of observing time available. Other resources which must not be exceeded include the amount of data which can be processed by the communications and ground systems, onboard tape recorder storage, and onboard command computer storage.

Figure III illustrates schematically the range of constraint timescales for HST. The interaction of so many constraints on varying timescales makes it impossible to identify any one dominant scheduling factor. Many of the constraints are periodic with different periods and phases. As a consequence there are generally several opportunities during a year to make a particular observation and a prime goal of HST scheduling is to make a near-optimal choice among these opportunities for as many observations as possible. This effort is complicated by the fact that the majority of the requested exposures have timing, grouping, repetition, or ordering constraints that couple very strongly with the time-dependent constraints of Figure III.

Constraints convey two types of information to the scheduler:

- *Feasibility constraints* (or "strict") constraints specify conditions or times when activities may or may not be scheduled. A few examples in the HST scheduling context are: provide a minimum of two months between an observation and a repeat observation on the same target or never schedule an observation when the Sun is within 50° of the target.

- *Preference constraints* specify quality judgments on scheduling conditions which are preferred but not required. These may be based on based on objective or subjective factors. In HST scheduling, for example, it is desirable to schedule at times which minimize scattered light from the bright limb of the Earth or maximize the chance of successfully acquiring guide stars.

It is important that both feasibility and preference information be considered *simultaneously* during schedule construction. Ignoring feasibility constraints can obviously lead to unimplementable schedules, but disregarding preference constraints (in order to simplify the problem) can lead to unacceptably suboptimal schedules. For this reason the concept of *suitability functions* (Johnston and Miller 1994) was developed by merging ideas from two well-studied frameworks, namely constraint satisfaction problems (CSPs) for expressing and manipulating feasibility constraints, and evidential reasoning techniques as a means to combine preference constraints.

Spike represents schedules in the format of a CSP which consists of a set of variables each with a domain of discrete values, and a set of constraints that limit the allowed values for each variable based on the assigned values of other variables. With each observation represented by a variable and the time of observation represented by the value of the variable, a CSP describes the telescope scheduling problem. (In this framework time is discretized, but the discretization can be made sufficiently fine, e.g. a week, a day, a orbit, etc.) CSPs are useful in a variety of applications and methods for studying them are widely published in mathematical, computer science and other fields (see Kumar 1992 for references). Most discussions of CSPs ignore preferences and consider only strict constraints. The Spike CSP does not make this simplification and considers preferences in the solution process.

Spike uses a scheduling method called *multistart stochastic repair* (Johnston and Minton 1994) which consists of the following steps:

1. Initial assignment - make an initial assignment ("guess") of observations to times based on heuristics discussed below. Such a schedule will generally have constraint violations and resource capacity overloads.

2. Repair - apply heuristic repair techniques to eliminate constraint violations until there are none left or a pre-established level of effort has been reaches (e.g. N iterations).

3. Deconflict - eliminate remaining constraint violations by removing activities with conflicts or relaxing constraints until a feasible schedule remains.

The heuristics employed by Spike are stochastic, so there is a benefit in repeating the above steps as often as time allows. The general strategy is to select the best of many runs, usually trying different guess and repair heuristics. Note also that this approach has the desirable "anytime" characteristic: at any point in the processing after the initial guess has been constructed a feasible (but not necessarily near-optimal) solution can be produced simply executing the deconflicting step. This is especially useful for quick response to schedule changes.

A more detailed discussion and analysis of the heuristics used in Spike is found in Johnston and Miller (1994) and Johnston and Minton (1994). The system for devising the initial assignments in Spike is described by Sponsler (1994).

IMPROVING THE PROCESS

To support Cycle 5 observations, the STScI is introducing a number of improvements. In previous cycles, only the RPSS validation software was available to the proposer - Transformation, Spike and other tools were run by STScI staff. With the advances in computer display and networking technology over the past few years it is now possible to make more STScI tools available to proposers and to graphically display the results (Bose 1995). Providing observers with these tools allows them to understand the impact of planning and scheduling decisions on the scientific goals of the program and it makes it easier to create correct observing proposals. This new system is called "RPS2" (second generation Remote Proposal Submission). In addition to providing greater insight to the planning and scheduling process, RPS2 is being designed to facilitate changes to proposals during execution. It is clear that scientifically valid changes to the observing program should be facilitated, not discouraged, by the planning and scheduling system. RPS2 will facilitate changes to proposals in several ways. First, it will show observers the current state of each observation (archived, executing, awaiting flight scheduling, etc.) so that the impact of a change can be readily determined. Second, RPS2 will show the effect of changes on related observations. Third, the syntax of the RPS2 file makes it easier to construct modular proposals, without compromising the ability to express complex scientific requirements.

Figure IV illustrates the information available for a proposal from the STScI Program Information World Wide Web (WWW) server (http://presto.stsci.edu/propinfo.html). The leftmost panel shows the initial page. The user can enter the proposal id (if known) or the last name of the Principal Investigator (PI) if the id is not known. Links to current HST short- and long-term schedule information are also found on this panel. The center panel shows information for a particular proposal and gives the PI name, proposal title and overall proposal status. Links to the proposal file and information about specific observations

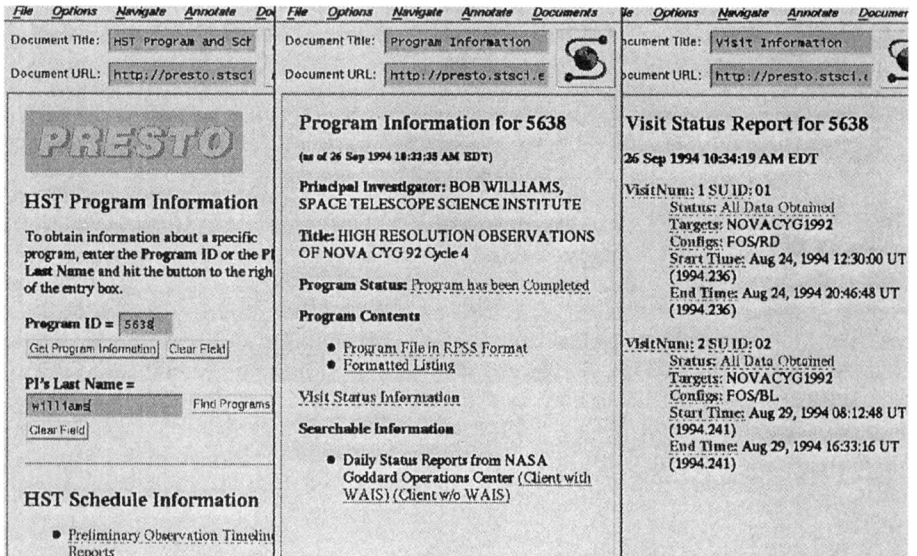

FIGURE IV Proposal information available via the World Wide Web.

(visits) are available. The rightmost panel shows the visit status information for the two visits in this proposal.

Another improvement in the planning and scheduling process relates to long-range planning. In Cycles 0-3, the STScI provided a long-range plan to observers which specified the predicted week of execution. This plan was unstable for two reasons. First, the proposal pool changed frequently, which caused changes in the planned execution date. Second, it was necessary to oversubscribe each week in the long-range plan to achieve a high level of efficiency. That is, since it was not possible to predict in advance which observations would be executed in a particular week (due to proposal changes and uncertainties in scheduling), allocating additional observations to each week was necessary to prevent unused HST observing time. Although many observations were not executed in their planned week, it was found that nearly all observations were executed within a few weeks of the planned date. For late Cycle 4 observations and beyond, observations will be commited to a broader "plan window" which is several weeks in duration. This will give proposers an accurate indication of when an observation will be executed without being overly sensitive to the vagaries of proposal changes and scheduling.

APPLICATION TO OTHER ASTRONOMICAL SCHEDULING PROBLEMS

Spike has been adapted to schedule a variety of astronomical scheduling problems (see below). Of these, ASCA and EUVE are in flight operations (in addition to HST), while several others are in the prototype or planning phases. The experience of customizing Spike for other types of problems has been actively sought during Spike development: each case provided feedback on the approach, and led to improvements from one version to the next.

- Hubble Space Telescope (HST) - Operational since Oct. 1989, HST launch Apr. 1990. Used for HST long-range planning at STScI.

- Extreme Ultraviolet Explorer (EUVE) - Spike operational since Apr. 1991, EUVE launch in Jun. 1992. Used for one-year scheduling of pointed observations. Run by Center for Extreme Ultraviolet Astrophysics, Univ. Calif., Berkeley.

- ASCA - Operational since Nov. 1992, flight operations began following launch in Feb. 1993. Spike is used for long-term scheduling. Joint Japan/US X-ray telescope mission run from the Institute of Space and Astronautical Sciences in Japan (Isobe et al. 1993)

- X-ray Timing Explorer (XTE) - Planned for use following launch in 1994. Spike would schedule both long-term and short-term. XTE will be run from the GSFC XTE Science Operations Center.

- Advanced X-Ray Astronomy Facility (AXAF) - Prototype developed 1990 as part of successful science operations center proposal. Under consideration for both science and mission scheduling.

- Roentgen Satellite (ROSAT) - Prototype developed 1992 for feasibility evaluation for operational X-ray satellite. Dual long-term/short-term scheduling mode. German Space Operations Center, Munich.

- International Ultraviolet Explorer (IUE) - Prototype developed 1988 for evaluation of Spike framework. Scheduling mode was 6-months of European half-shifts, optimized for target coverage (Johnston 1988a).

- Ground-Based - Prototypes developed in 1988-1992 to demonstrate feasibility. Dual mode: long-term scheduling for night allocation, short-term scheduling for exposure scheduling within a night. Telescopes included ESO (European Southern Observatory) and CFHT (Canada-France-Hawaii Telescope), as well as hypothetical Automatic Photometric Telescope (APT). Feasibility of coordinated ground- and space-based scheduling has also been demonstrated (Johnston 1988b).

The adaptation of Spike for these problems demonstrates the flexibility of the framework. Spike was designed so that new tasks and constraints can be defined without changing the basic framework. For ASCA and XTE, Spike is operated in a hierarchical manner, with long-term scheduling first allocating observations to weeks much as they are for the HST problem (and with similar types of long-term constraints and preferences). Then each week is scheduled in detail, subject to the detailed minute-by-minute constraints of low earth orbit operation. The major changes required to implement short-term scheduling were:

1. A new type of task that can have variable duration depending on when it is scheduled, and which can be interrupted and resumed when targets are occulted by the Earth or the satellite is in the radiation belt (i.e. task preemption).

2. New classes of short-term scheduling constraints which more precisely model target occultation, star tracker occultation, ground station passes, entry into high radiation regions, maneuver and setup times between targets, etc.

3. An interface between different hierarchical levels, by which a long-term schedule constrains times for short-term scheduling and conversely.

4. A post-processor which examines short-term schedules for opportunities to extend task durations and thus utilize any remaining small gaps in the schedule to increase efficiency.

Most of the effort required to apply Spike to the new problems was limited to the specific domain modelling necessary, which typically involves computation related to the geometry of the satellite, Sun, target, and Earth. These problems can be expected to differ from one osbservatory to another, and it is not surprising that different models are required.

EUVE is unusual in that it makes long (2-3 day) observations, in contrast to HST and ASCA which typically make numerous short (15-40 minute) observations. As a consequence, EUVE is schedulable over year-long intervals without breaking the schedule into hierarchical levels. One of the more interesting results from a comparison of search algorithms for scheduling EUVE was that the Spike repair-based methods gained an extra 20 days of observing time in a year, when compared to the best incremental scheduling approach.

ACKNOWLEDGMENTS

Many people at the STScI have contributed to the development and implementation of the systems described here.

REFERENCES

Adorf, H.-M. 1990, Space Telescope European Coordinating Facility Newsletter, 13, 12-15.

Asson, D. 1995, Proc. of the SPIE AeroSense '95 Conference, submitted.

Bose, A. 1995, Proc. of the SPIE AeroSense '95 Conference, submitted.

Drummond, M., Swanson, K., and Bresina, J. 1994, Intelligent Scheduling, ed. M. Zweben and M. Fox (San Franciso: Morgan Kaufmann), 341-370.

Genet, R., Genet, D., Talent, D., Drummond, M., Hine, B., Boyd, L., Trueblood, M. 1991, in Robotic Observatories in the 1990's, ed. A. Filippenko, ASP.

Gerb, A. 1991, Proceedings of the Goddard Conference on Space Applications of Artificial Intelligence, NASA CP 3110, 283-295.

Isobe, T., Johnston, M., Morgan, E. and Clark, G. 1993, Proc. 2nd Conf. Astron. Data Analysis and Software Systems, ASP.

Jackson, R., Johnston, M., Miller, et al. 1988, Proc. of the Goddard Conference on Space Applications of Artificial Intelligence, NASA CP 3009, 197-212.

Johnston, M. 1988a, Proc. ESA Workshop on AI Applications for Space Projects, (Noordwijk: ESTED), 5-9.

Johnston, M. 1988b, Proc. ESO Conf. on Very Large Telescopes and their Instrumentation, (Garching: ESO), 1273-1282.

Johnston, M. and Miller, G. 1994, Intelligent Scheduling, ed. M. Zweben and M. Fox (San Franciso: Morgan Kaufmann), 391-422.

Johnston, M. and Minton, S. 1994, Intelligent Scheduling, ed. M. Zweben and M. Fox (San Franciso: Morgan Kaufmann), 257-289.

Kumar, V. 1992, Aritificial Intelligence Magazine, 13, 32-44.

Miller, G. 1989, in Knowledge Based Systems in Astronomy, ed. F. Murtagh and A. Heck (Berlin: Springer Verlag), 5-32.

Sponsler, J. 1994, Proc. Goddard Conference on Space Applications of Artificial Intelligence, in press.

Advances in Autonomous Operations for the *EUVE* Science Payload and Spacecraft

T. Morgan and R. F. Malina

Center for EUV Astrophysics, 2150 Kittredge St., University of California, Berkeley, CA 94720-5030; tomm@cea.berkeley.edu and rmalina@cea.berkeley.edu

Abstract. We examine the use of science payload and spacecraft autonomy and artificial intelligence (AI) in ground systems operations to reduce staffing requirements and lower overall mission operations and data analysis (MO&DA) costs for the *Extreme Ultraviolet Explorer (EUVE)* mission. Applying the robotic telescope paradigm to the *EUVE* observatory, we have identified a three phase evolution toward the eventual goal of full-loop, scientist-to-observatory automation. The study focuses on streamlining the current, human-intensive support structure. The streamlining, implemented in phases, will result in an automated "hands-off" operations mode that poses no threat to the mission's requirements. The transition will rely heavily on the automated, smart safing modes built into the *EUVE* Science Payload and Explorer Platform (EP) flight software and innovative approaches to space communications. This paper outlines the three proposed implementation phases, the portions of the robotic telescope model that can be directly transferred to that phase of *EUVE* automation, as well as projected staffing reductions for each phase of implementation.

1. INTRODUCTION

The *EUVE* satellite is the first observatory dedicated to scientific study of the extreme ultraviolet (EUV) region of the electromagnetic spectrum. *EUVE* is a two-component observatory, comprising a science payload and a supporting spacecraft. *EUVE* launched June 7, 1992, from Cape Canaveral Air Force Station at Kennedy Space Center, FL. In August 1994 *EUVE* completed its primary science objectives. It has performed flawlessly with only one spacecraft system failure in two years of operation, and the spacecraft "safehold" mode has never been exercised. *EUVE* is now 18 months from the end of its prime mission, and the University of California at Berkeley (UCB) is compelled to look at radical, new, low-cost approaches for operating the observatory or faces having the satellite turned off for budgetary reasons at NASA. The model proposed a decade ago by the ground-based community and the pioneers of robotic telescopes offers an attractive, low-cost model applicable to present and future orbiting missions. The ten years of research and development that have led to the technological advances of the robust and productive robotic telescopes operating today have resulted in many techniques that are transferable to orbiting observatories. A cultural bridge must be crossed if NASA is to be convinced of this approach on any large scale, but that bridge is largely one of convention. Traditionally, space missions are conservative by nature, using only time-hardened technology, often dating a decade or two behind current innovative capabilities. The scenario of full-loop, scientist-to-observatory automation is technologically possible on a large enough scale to support orbiting missions because of current advances in AI technology and declining prices of hardware associated with distributed architectures. UCB proposes to use *EUVE* as a "testbed" to validate the emerging, innovative technologies as applied to in-orbit observatories.

2. MARGINS

The science instruments aboard *EUVE* were designed, built, and tested at Space Sciences Laboratory (SSL) on the Berkeley campus. The payload design occurred in parallel with the evaluation of perspective spacecraft bus designs. A final decision was made in 1985 to use the Fairchild Aerospace Multimission Modular Spacecraft, later known as the Explorer Platform (EP). The years of independent development, without a specified spacecraft bus, precipitated the parallel, redundant margins necessary for the scope of automation UCB is implementing. Instrument scientists and engineers at UCB faced large "unknowns" in the spacecraft-to-payload interface. The lack of a firm spacecraft design forced the *EUVE* payload to handle virtually all functions internally and independently. The payload requires from the spacecraft only a reference clock, a single 16-bit serial command input, 28 volts of power, and control of pyrotechnic devices for uncovering the telescope apertures. However, with the uncertainty of a spacecraft design, the stand-alone payload design allowed UCB to proceed in its development efforts while NASA contracted for a spacecraft (Marchant, 1993). The end product was an *EUVE* observatory with independent and redundant safing systems and an extremely simple and problem-free interface.

3. PRIME MISSION OPERATIONS

A unique mission operations structure supports the *EUVE* observatory. NASA Goddard Space Flight Center (GSFC) in Greenbelt, MA, manages the project. GSFC performs all space craft support functions from orbital determination and commanding to health and safety monitoring of the spacecraft and level-zero processing of the science data. The *EUVE* Science Operations Center (ESOC), which controls the science payload, is located at the Center for Extreme Ultraviolet Astrophysics (CEA) at UCB. The ESOC is the primary interface between UCB and GSFC. UCB provides a multitude of science processing and support functions to prepare data as well as carry out science planning and monitor health and safety of the *EUVE* science payload. The ESOC and GSFC exchange three types of data:

- science data, captured by the instrument,
- engineering data, health and safety status of the observatory, and
- navigation data, spacecraft position and attitude.

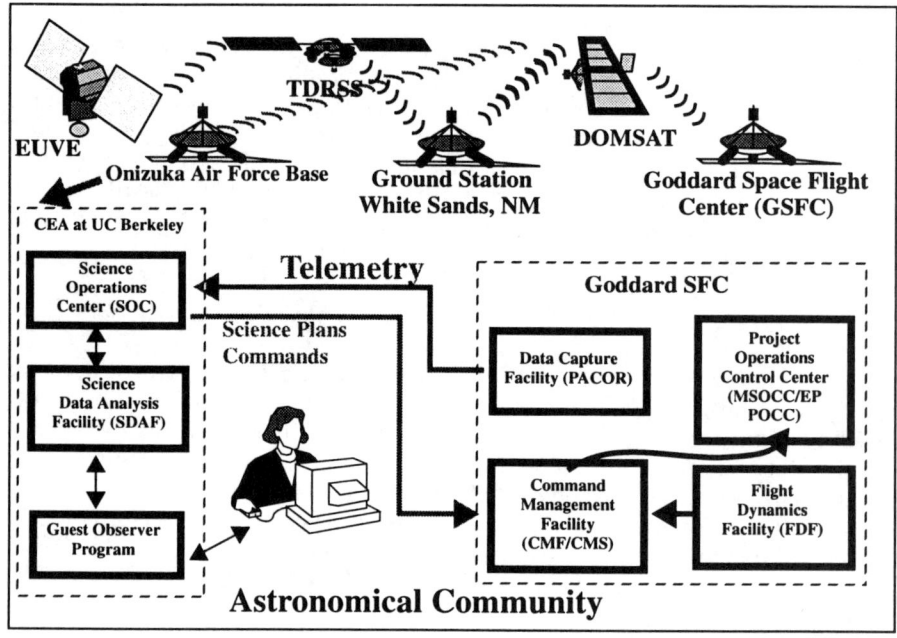

Figure 1. Diagram of current data flow and support functions.

Spacecraft and instrument health and safety monitoring is the most labor intensive of the support functions. Both GSFC and UCB conduct health and safety monitoring on a 24 hour per day, 365 day per year basis. Spacecraft/payload controllers receive health and safety (engineering) telemetry every 90 minutes. Data are transferred from the spacecraft to UCB via TDRSS-White Sands-DOMSAT-GSFC relay (Fig. 1) utilizing a series of secure, high-speed communication networks (NASA document, 1989).

4. SIGNS OF A NEW MO&DA CLIMATE

In the February 1993 report *Technology for America's Economic Growth, A New Direction to Build Economic Strength,* the Clinton Administration formally recognized technology as the engine of economic growth and acknowledged scientific advances as the foundation upon which technical progress is built. The report states that "all laboratories managed by DOE, DOD, and NASA that can make a productive contribution to the civilian economy will be reviewed with the aim of devoting at least 10–20% of their budgets to research and development partnerships with industry." The report also stated that "all federal support for technology development will be reviewed to ensure research priorities address industry needs, and every federal technology program will be regularly evaluated to determine if it should remain part of the national program" (Huntress, 1994). NASA, along with many other federal agencies, is currently working with the Administration's Office of Science and Technology Policy, the Office of Management and Budget, and Congress to promote technology as a means for economic growth. NASA's Office of Space Science (OSS) is supporting this national technology thrust by:

- exercising world leadership in basic science, mathematics, and engineering through sponsorship of scientifically and technically challenging space science missions;
- supporting the development and infusion of state-of-the-art technologies into its science missions; and
- establishing partnerships with US industry to transfer these technologies to the private sector, thereby providing tangible benefits to the economy.

"The fundamental mission of OSS is to further our understanding of the universe, its origin, and the stellar and solar system, and to direct this understanding to practical applications where appropriate. To achieve this mission, OSS must foster the development of new technologies to continually improve scientific capabilities, and transfer science and technological advances to the public and private sector to assure US scientific and technical leadership and benefit quality of life for all" (Huntress, 1994).

The Administration's and NASA's renewed focus on technology found a welcome audience at CEA. The 12 years of $EUVE$'s design and fabrication were driven by innovative, high-tech approaches to complex instrument design. Examples range from the development of detectors never utilized in an orbiting mission to the SUN Microsystems distributed workstations architecture on which the $EUVE$ ground systems were designed and operate. The catalyst that unified UCB's and NASA's interest in low-cost innovative technology was the relationship developed at NASA headquarters (HQ) between the Office of Advanced Concepts and Technology (Code C) and OSS.

5. CODE C ASSISTANCE

In December 1992, UCB contacted NASA HQ Code C to seek assistance in lowering the overall MO&DA costs for $EUVE$. Code C coordinated a systematic response to the CEA strategic plan to lower operations costs through NASA Ames Research Center (ARC) and the Jet Propulsion Laboratories (JPL). ARC and JPL were enthusiastic about assisting the $EUVE$ project. After a series of meetings, it was determined that the ARC, JPL, and CEA collaboration would focus on the UCB role of telemetry monitoring, in particular the checking of science payload health and status.

As previously stated, the current baseline method of checking the instrument health and status is very labor intensive, with the majority of the controllers' time on shift spent watching numerical values fluctuate between limits. The collaboration identified this area as one to which automation could be applied with high investment return and set out to identify an appropriate technology.

JPL and ARC maintain an archive of operations software, which they made available to UCB for evaluation purposes. A parallel activity was initiated at UCB to survey commercial off-the-shelf (COTS) products as well as other NASA center software. The evaluation team was given three months to make a recommendation. The team chose a low-cost COTS package called RTworks from a Sunnyvale, CA, based firm, Talarian Corp. The requirements, drivers, and complete evaluations of the surveyed software are available via Mosaic (http://cea-ftp.cea.berkeley.edu) on the CEA home page. The summary of the findings describes RTworks as the only package available at the time that was flexible, well documented, and included good software support for customers.

6. POLIDAN STUDY

On July 1, 1993, Dr. Guenter Riegler of NASA HQ Code SZ commissioned a study to determine the feasibility of extending the *EUVE* mission beyond January 1996. Dr. Ron Polidan headed the study team, which released its findings in February 1994. The Polidan study defined the focus of the on-going CEA research and provided a logical framework and plan for the implementation of the proposed low-cost, innovative technologies. The study contained three decisive, phased scenarios for extending the *EUVE* mission:

1. Reduce operations to one shift in a phased approach (three shifts, to two shifts, to one shift) for a stand-alone, low-risk solution to Code S operations cost reduction.

2. Reduce to one-shift operations as soon as possible and redirect savings to develop a more automated control center to reduce *EUVE* operations cost and provide an innovative solution that can be used as a model for future mission developments.

3. Transfer spacecraft operations to UCB, and use direct-to-ground (DTG) communications as soon as feasible to reduce cost and demonstrate portability and robustness of the workstation-based operations design.

The study recommended following a three phased approach, "First, Implementing option number 1 to reduce costs. The[n] implement option number 2 (after ensuring it is cost effective) since it would provide for continuing technology playback to NASA during the transition as well as into the extended mission. It also maintains a high level of science return and allows for mission operations to continue at GSFC which ensures a higher level of support for anomaly resolution. Finally, in the later years of the mission, implement option 3 to demonstrate feasibility of handling over operations to the user [*sic*]" (Bruegman, 1994).

The overall UCB implementation philosophy attempted to remain as orthodox to the Polidan recommendation as possible. The only alteration made was to the order in which steps 1 and 2 were executed. UCB determined that the risk associated with executing option 1 without having option 2 in place was too high to remain safely within the *EUVE* mission requirements (as specified in the Mission Requirements Documentation; NASA document, 1989).

7. PHASE 1: LOW-COST TELEMETRY MONITORING IMPLEMENTATION

The Polidan study concluded that, "Reducing operations to one shift in a phased approach, (three shifts, to two shifts, to one shift) is a stand-alone, low-risk solution to Code S operations cost reduction," and recommended that *EUVE* investigate methods "to reduce to one-shift operations as soon as possible and redirect savings into developing a more automated control center." In the spring of 1994, UCB performed the survey of NASA-developed and COTS software that would, once selected, be implemented in the ESOC to allow the phased staff reduction as well as ensure health and safety of the spacecraft.

The implementation task was a formidable one. It involved converting the current mode of operations based on human-centered intelligence into lines of knowledge-based rules, which are fed into an inference engine for interpretation and inferred action. The conversion method used proved very effective. It entailed having the payload controllers, who actually performed the daily monitoring, represent their domain experience and function in an IBM flowchart (Fig. 2). Prior to selecting this format, CEA performed a survey of the available representations, digraphs, fault trees, and flowcharts. The IBM flowcharts proved easiest to use because of prior staff experience and familiarity. The duty scientists then reviewed the flowcharts. Most flowchart representations required three or four iterations before meeting the requirements for sign-off. Once signed off, the knowledge representations went to a knowledge/software engineer. The knowledge engineer met with the payload controller who constructed the knowledge representation for a walk-through, preliminary explanation. A draft version of the code was written and incorporated into the RTworks architecture existing in the testbed. The responsibility then fell back onto the domain expert/payload controller to test the automated knowledge base and evaluate its performance versus the nominal response of a controller.

This method has proven successful in a remote engineering scenario as well. Researchers at ARC in the Augmented Monitoring and Diagnosis Application group acquired a duty scientist approved version of the global health and status knowledge representation via the internet and converted it into a knowledge base and rules that was FTPed back to CEA and appeared, as ARC coded it, in a July 1 software release.

As of August 15, 1994, testing began in a parallel operations mode. The automated monitoring system operates in parallel with the human-centered intelligence system. After two months of testing, the ESOC will be staffed for 14 hours by a robust and stable AI based, automated, telemetry monitoring system linked to an autonomous paging system. Human staffing will move to one, 10 hour payload controller shift, with the other 14 hours being monitored only by the AI software. In the event of an anomalous condition, the broken rule will cause the inference engine to trigger a series of pages to anomaly response coordinators (humans) who will respond and resolve the anomaly or obtain further support from experts. In the future, as the RTworks-based system stabilizes, the duration of the single shift will evolve toward "zero shifts," leaving the health and safety of the instrument monitored by software alone. After a final move to a zero-shifts situation, the science payload engineer will conduct periodic engineering analyses to detect any degradations or changes in the instrument that are

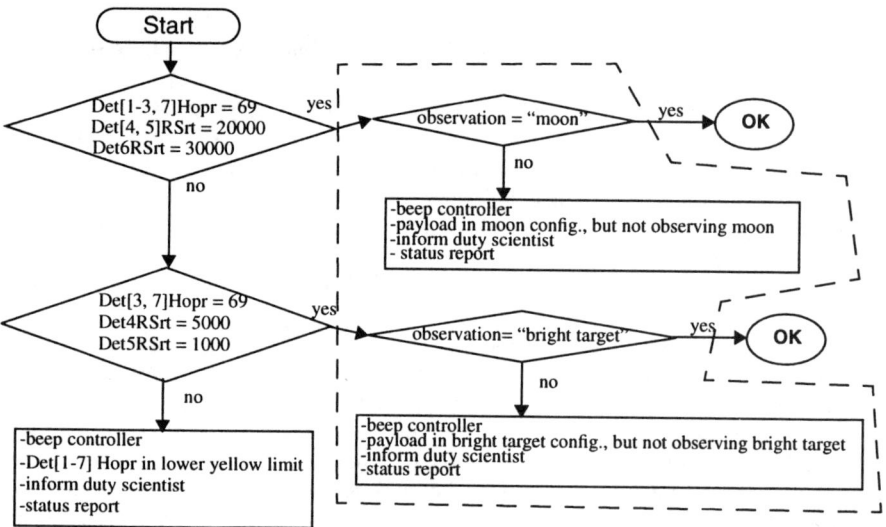

Figure 2. Special Observations Flowchart: This flowchart satisfies the requirements for the September and October 1994 3-to-1 shift simulation only. Note: "observation" is a parameter external to the telemetry stream.

not yet detectable by the AI software. This will require automation of all tasks currently moved to the human-tended shift, as well as all those tasks currently moved off the shift like trending and engineering analysis. It is anticipated that the move from one to zero shifts will require a much larger degree of automation than already accomplished in the move from three to one human-tended shifts.

8. PHASE 2: INNOVATIVE, LOW-COST SPACE COMMUNICATIONS

In parallel with the telemetry monitoring development, a study was initiated to look at innovative approaches to space communications. The Low Earth Orbit Demonstrator (LEO-D) at JPL, commissioned under Charles T. Force of NASA HQ Code O, initiated contact with *EUVE* some 12 months ago. The LEO-D

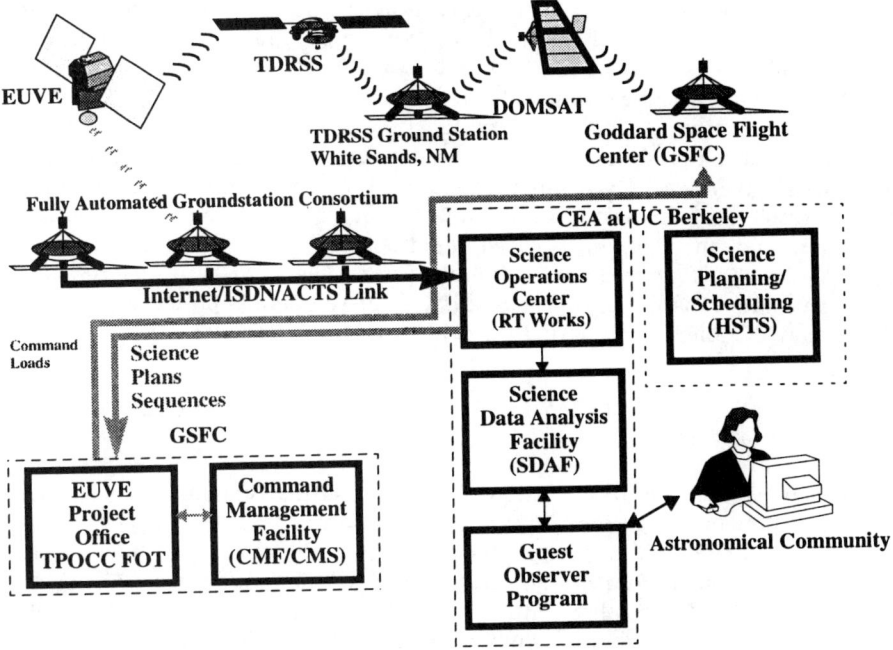

Figure 3. TDRSS hybrid data flow and support functions.

project proposes to use low-cost weather satellite ground stations to receive low-earth-orbit science-satellite data. A successful demonstration of the LEO-D project promises great strides toward the goal of full-loop, satellite-to-scientist automation. *EUVE* utilizes TDRSS and a series of support groups at GSFC for communications with the ground. TDRSS is largely over subscribed and very costly. The LEO-D model would allow a mission to autonomously acquire orbital determination information, lock on the spacecraft, receive real-time or tape recorder telemetry, execute level-zero processing and package and ship the data to a user. A strategy under development would use completely automated, low-cost, weather satellite ground stations to monitor the return link from the observatory and would maintain TDRSS for the forward link (Fig. 3). This combination provides the capability for health and safety checks of the observatory during any real-time ground contact and maintains a virtual back-up communications system, using TDRSS, which ensures bi-directional communications between *EUVE* and the ground for 20–30 minutes. Health and safety concerns detected by the automated ground stations would be assigned a "severity" level, with the highest level of risk initiating the activation of a human response team that could use TDRSS for diagnostics and rescue of the observatory if necessary.

In order to download data to receive-only ground stations, a satellite will require a patch of the ground ephemerides in advance to initiate data download when in range of a ground station. The vehicle aquisition would require an uplinked command sequence determining the location of the satellite and ground station into on-board computer memory. The commands that will schedule a dump will require an available and functioning ground station. Spacecraft ephemeris tables based on elements found at the North American Air Defense (NORAD) element archive will be uploaded during any real-time TDRSS contact to provide the necessary ground station location information.

This model could be augmented to allow for health and safety monitoring at the site with only science data being transported to users. The LEO-D program only addresses automating the return link in the equation. Science planning, sequencing, and all command related functions still depend on TDRSS and associated support functions for communications. The LEO-D has successfully demonstrated its capabilities on the SAMPEX mission and is preparing to carry out a similar demonstration with *EUVE* in the late fall of 1994.

9. PHASE 3: FULL LOOP, SATELLITE-TO-SCIENTIST AUTOMATION

The third phase will be an extension of the foundation offered by phases one and two, as previously outlined. Phase three will extend the use of AI automation to reduce ESOC staffing from one shift to zero shifts and will enhance the LEO-D model to incorporate forward and return link capabilities. All observatory telemetry will be relayed directly to the ground via a low-cost ground station, and all commands and software maintenance will be uploaded via a secure US located low-cost ground station.

Upon a successful 3-to-1 shift transition and a six month stabilization period, CEA will eliminate all staffing requirements for the payload controller duties. The rationale behind the transition is based on the results of mean time to failure (MTF) statistics associated with the 3-to-1 shift transition stability and the requirement to have the AI software operating at all times. Nominally, the AI software operates while payload controllers are present and when they are not. The only variation the software registers is the change in method for warning the payload controller; i.e., should the software call someone via telephone and beeper or simply sound an on-screen alarm in the ESOC? Once an acceptable MTF value has been attained, the software development effort will be frozen. All anomaly warnings or alarms the software detects will be channeled to the remote pager system, and all staff associated with routine payload monitoring activities will be removed. The eventual goal of Phase 3 is to have all health and safety monitoring functions performed at a grid of small, low-cost ground stations (Fig. 4). All telemetry would be processed at the ground station to level-zero and then be sent directly to the PI or guest observer. Commanding functions are performed via a secure US located ground station, based on the GSFC developed TPOCC software. Phase 3 relys only on a PI or science planner: (1) to prioritize target acquisition requests and send the science plan to a central server that would schedule ground contact necessary to support target requests,[1] (2) to maintain the network systems, and (3) to mobilize the necessary expertise for anomaly resolution or someone with significant spacecraft engineering expertise who is maintained through GSFC on an as-needed basis. GSFC spacecraft engineers have accumulated a two-year history in daily routine and contingency procedures for safe operation of both ground and in-orbit assets of *EUVE*. This relationship would be maintained by a low-level support contract. The ground systems architecture is being developed in a distributed manner, with remote response capabilities a primary concern. The network management and remote anomaly response tools are driving the construction effort of future *EUVE* ground systems for remote anomaly response. The strategic goal for the ground systems is to have the capability of fully remote response to any observatory anomaly. This up-front design would reduce to nearly zero the maintenance cost associated with engineering response to observatory problems.

The Phase 3 communications systems enhancements would be evolutions in capabilities of the existing hardware with a primary focus on the inclusion of forward link capabilities. The existing Phase 2 architecture maintains the return link capability. All observatory telemetry is delivered via DTG. Phase 3 would extend the capabilities to include forward link as well. Commanding, flight software maintenance and anomaly resolution efforts would be relayed via secure, low-cost ground stations. Hardware modification would only be necessary for a single, command-designated ground station.

10. ASSOCIATED SUPPORT FUNCTIONS

The full-loop, satellite-to-scientist, automated *EUVE* scenario proposed here is a simplified model. The numerous support functions related to communications, orbital determination, data retrieval, command generation and validation, and spacecraft maintenance will be performed locally at the ground station site, GSFC, or a PI science center.

11. CONCLUSION

The future of full-loop, satellite-to-scientist automation is near for small, scientific satellites. Automatic photometric telescopes (APTs) are proof no mechanical or technical limitations to automation exist that cannot be overcome; the limitations to automation of space missions are cultural. The experience with the ALEXIS mission provides one prototype for future low-cost operations; *EUVE* will offer a testbed for extending this technology to a sophisticated observatory. The unification of a decade of experience in ground-based automation with a space mission possessing systems redundancy produces a viable testbed for the development of promising new approaches to some old and costly problems. CEA proposes to use the APT model and the UCB vision to produce a viable low-cost operations model that will provide insight into missions for today

---

[1] The human factor remains in the subjective selection of target proposals. The labor-intensive nature of science planning is reduced by the constraints of model-based scheduling systems like HSTS under development by Nicola Muscettola at ARC.

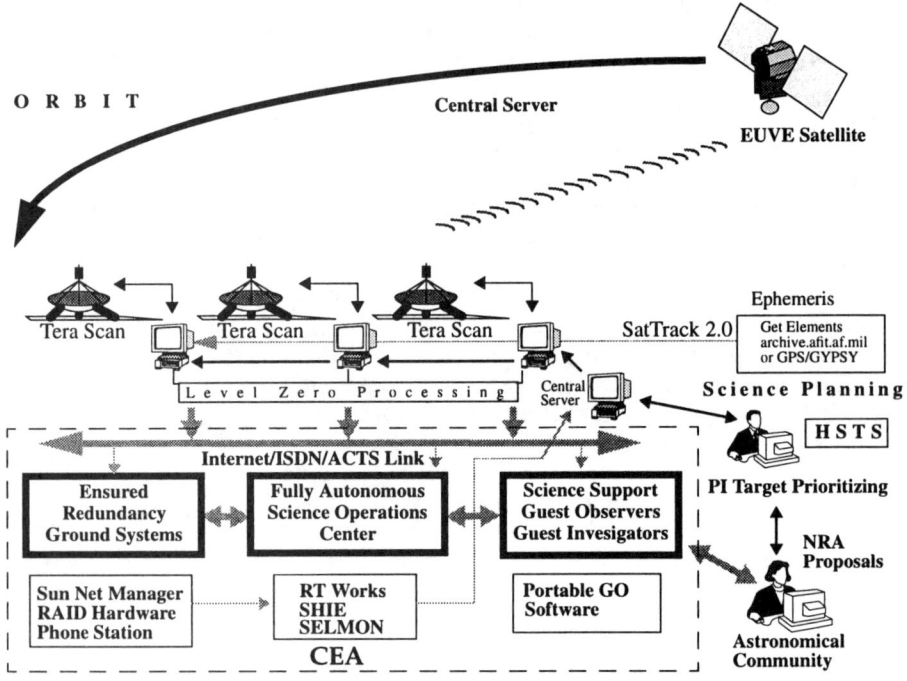

Figure 4. Fully autonomous ground stations consortium.

and tomorrow. Only with earnest cooperation between the ground-based and orbiting communities can this goal be attained.

Acknowledgments. We acknowledge development assistance from W. Marchant, N. Golshan, L. Wong, F. Girouard, A. Hopkins, P. Eastham, D. Iversen, and A. Pattersen-Hine. We thank the *EUVE* science team, the *EUVE* project technologist Peter Hughes, and AI researchers David Korsmeyer and Peter Friedland for their help and support. This research has been supported by NASA contract NAS5-29298 and NASA Ames grant NCC2838.

References

Bruegman, O., & Douglas, F. 1994, Study of Low Cost Operations Concepts for *EUVE* Mission Operations, p. ii–vi

Huntress, W. T., & Reck, G. M. 1994, Office of Space Science Integrated Technology Strategy, p. 1–5

Marchant, M. 1993, *JBIS* 46, 353

NASA Document 408-07056, 1989, Explorer Platform/Extreme Ultraviolet Explorer, Systems and Operations Requirements Document

The South Pole Infrared Explorer (SPIREX): Near Infrared Astronomy at the South Pole

Bernard J. Rauscher, Mark Hereld, Hiên Nguyên, and Scott Severson

The University of Chicago, Department of Astronomy and Astrophysics, 5640 South Ellis Avenue, Chicago, IL 60637

Abstract. During the 1993-1994 austral summer, we deployed the first generation South Pole Infrared Explorer (SPIREX) to the South Pole. Over the next several years, we will deploy a series of spectrometers, imagers, and telescopes to the South Pole. The SPIREX project's goal is to exploit the low thermal background at the South Pole for near infrared observations. In coming years, we expect to map a substantial area of the sky to cosmologically significant depths. Scientific goals include studying the origins of galaxies and stars.

The current SPIREX system consists of a $60 cm$ telescope, a near infrared spectrometer and imager, associated supports systems, computers, and software. In future years, we will deploy larger format cameras and larger telescopes. The harsh climate at the South Pole dictates that all of SPIREX's systems be designed for remote operation. Remote operation is also necessary for the highly automated deep surveys we plan to begin next year.

We have been monitoring the near infrared background since April 1994. Preliminary analysis of these data indicates that the near infrared sky is significantly darker at the South Pole than at warm sites. This paper outlines the background measurement experiment. A future paper will provide quantitative results and analysis.

In July 1994, SPIREX observed Comet Shoemaker-Levy 9's collision with Jupiter. SPIREX was unique among earth based sites in having a nearly uninterrupted view of Jupiter during the collision period. SPIREX observed 16 impacts and obtained baseline coverage between impacts.

1. Introduction

We deployed the first generation South Pole Infrared Explorer (SPIREX) to the earth's South Pole between December 1993 and February 1994. This paper describes the SPIREX project with an emphasis on what we have done in the past year and what is occurring at the South Pole this austral winter.

The discussion is broken into 7 sections. Section 1 is this introduction. Section 2 briefly describes the advantages of the South Pole as a near infrared observing site. Section 3 describes SPIREX's hardware and software. Section 4 discusses SPIREX's goals this austral winter. Section 5 discusses our accomplishments to date. Section 6 discusses our goals for the coming year. Section 7 is a summary.

2. Advantages of the South Pole

The South Pole's principle advantage as a near infrared observing site is it's very low thermal background (Hereld 1994). At wavelengths shorter than $2.27\mu m$, the background is dominated by chemical reactions in the upper atmosphere, a main component of which is OH. OH airglow affects any ground based site. However, at wavelengths longer than $2.27\mu m$, thermal emission from the atmosphere and telescope dominates.

We have calculated that by moving the observatory to the South Pole, both the telescope and atmosphere are cooled, and the background in the K_{dark} band $(2.27 - 2.45\mu m)$ is reduced by a factor of 220 compared to a warm telescope.[1] In K_{dark} band, the atmosphere is transparent, and the warm telescope suffers a factor of 15 penalty in signal-to-noise-ratio. If all else is held equal, the difference in the noise translates to the following gains for background-limited observations:[2]

- source flux brighter by 2.9 magnitudes,
- telescope diameter larger by a factor of 15, or
- integration time longer by a factor of 220.

Based on these calculations, a well baffled telescope at the South Pole is nearly limited by zodiacal light, and capable of competing on an equal time basis with a much larger telescope at a warm site.

3. SPIREX Hardware and Software

SPIREX currently sits on top the Center for Astrophysical Research in Antarctica's (CARA) Antarctic Submillimeter Telescope and Remote Observatory (AST/RO) building. Next austral summer, it will move to a dedicated two story tall telescope tower adjacent to the new CARA/AMANDA lab building.

Figure 1 shows SPIREX as it existed in February 1994. The system consists of a low emissivity infrared optimized 60cm telescope, a flexible near infrared imager and spectrometer, a tent to shield the system from wind and snow, heater circuits, control computers, and software. We will first discuss the telescope, then the near infrared camera, and finally other aspects of the system.

3.1. Telescope

SPIREX's 60cm telescope (Figure 2) is a 3 mirror bent Cassegrain design that forms an f/10 beam at the focal plane. The telescope was designed to achieve

---

[1] The warm telescope is assumed to operate in K band $(2.0 - 2.4\mu m)$. The relative advantage is greater if the warm telescope is operated in K_{dark} band.

[2] The comparisons give what a warm telescope would require to compete with a telescope at the South Pole.

Figure 1. SPIREX on the roof of CARA's AST/RO building. Hiên Nguyêñ (foreground right), Fred Mrozek (background left), and Bernie Rauscher (background right) are shown pulling the "baby buggy cover" over the instrument. The "baby buggy cover" provides a windbreak for working on the instrument and shields the instrument when it is not in use. The telescope is rotated so that the tube is behind the mount. The near infrared instrument, GRIM, is enclosed in the insulating foam box in the center of the picture.

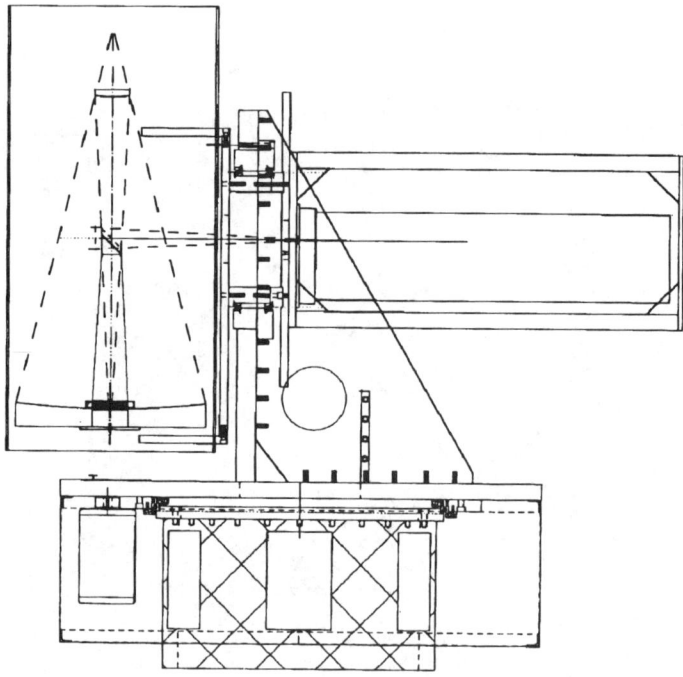

Figure 2. SPIREX's 60cm telescope is a 3 mirror bent Cassegrain design on an altitude-azimuth mount. It forms an f/10 beam at the focal plane. Specific infrared enhancements include using silver mirrors throughout and an undersized secondary. In this figure, the near infrared instrument (GRIM) is represented schematically by the rectangle to the right of the altitude bearing.

5% total emissivity. Specific infrared adaptations include using silver mirrors throughout and undersizing the secondary. The telescope has an altitude-azimuth mount. At the South Pole, this becomes an equatorial mount, and no instrument rotator is required. The axis are friction driven by a pair of Pacific Scientific stepper motors. There are heaters to melt any ice that may accumulate on the drive axis, or in the bearings. Important telescope sub-assemblies were extensively tested at dry ice temperature before being shipped to the South Pole.

The telescope control computer is a Dell PC running a customized version of the commercially available Sky Probe 1000 telescope control program. We are writing software that will allow the Dell PC to run a minimally modified version of the Apache Point Observatory (APO) telescope controller. The Apache Point software was designed explicitly for remote operation.

Figure 3. The Near Infrared Grism Spectrometer and Imager mounted on the Apache Point Observatory 3.5 m telescope. The instrument measures approximately $1.3m \times .6m \times .6m$. The cables entering from the left are two optical fibers, a serial cable, and an A.C. power cable. The electronics box closest to the telescope contains the clocking and readout electronics. The box farthest from the telescope contains stepper motor drivers and a thermometry circuit.

3.2. Camera

SPIREX's camera is the Near Infrared Grism Spectrometer and Imager (GRIM, Figure 3). GRIM was designed so that all of it's functions could be run remotely. It was extensively tested at APO between July and October 1992. GRIM incorporates an array of cold slits, a collimated beam section, a focuser, a grism for low resolution spectroscopy, a large selection of wide and narrow band filters, and 3 imaging scales (1, 2, and 4 $arcsec\ pixel^{-1}$). GRIM's detector is a Rockwell Science Center 128×128 NICMOS2 array sensitive from $1 - 2.5\mu m$.

The testing at APO included scientific observations by a number of observers at several institutions. This extensive testing allowed us to develop data acquisition and analysis strategies to best utilize GRIM's capabilities.

GRIM's graphical user interface (Figure 4) was designed to allow the user to control all aspects of the camera from the heated control room. It is very

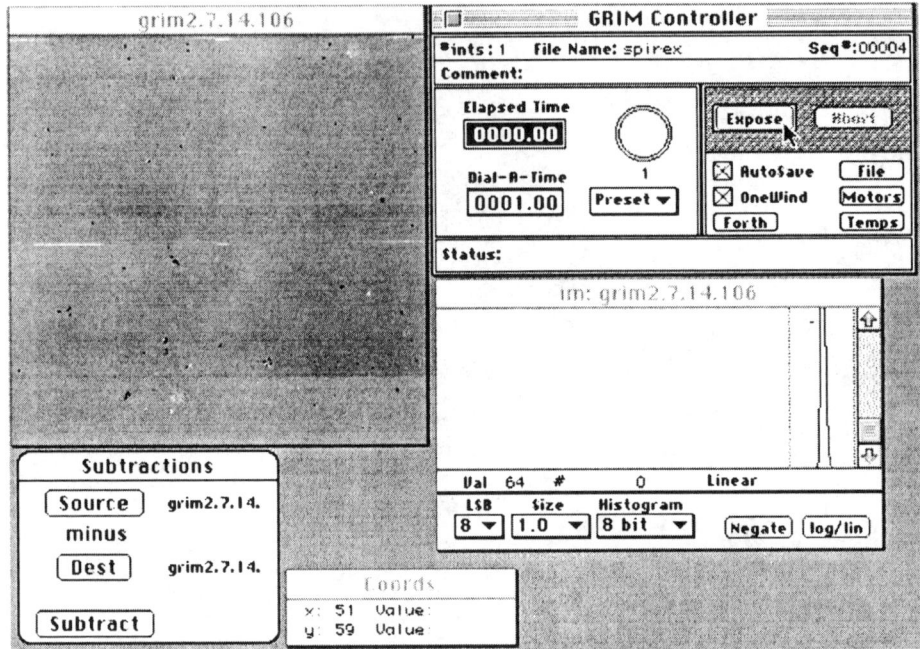

Figure 4. GRIM's graphical user interface allows the user to control all aspects of GRIM from the heated control room. The "GRIM Controller" window's (upper right) features include dialing in exposure times, multiple exposure modes, and changing the instrument's configuration (filters, magnifications, etc.). The "Histogram" window (lower right) interactively changes the image's appearance, but not it's content. The IMAGE window (upper left) displays the data. In this case, a 256 × 256 NICMOS3 frame taken with GRIM's sister instrument, GRIM II, is displayed. The SUBTRACT window subtracts images.

close to the APO user interface used by GRIM's sister instrument, GRIM II. Loewenstein et al. (1992) discuss APO software.

4. 1994 Austral Winter Goals

Our goals this winter are:

- measure the near infrared background,
- test SPIREX's mechanical subsystems,
- test operations at the South Pole,
- test interaction with isolated collaborators,
- begin taking deep K_{dark} band exposures, and
- observe Comet Shoemaker-Levy 9's collision with Jupiter.

Our highest priority this winter is measuring the background. These measurements have been ongoing since April 1994. The data now include over 3600 individual exposures. These data include spectra, K_{dark} band, J ($\lambda_0 = 1.25\mu m, \Delta\lambda = .3\mu m$), H ($\lambda_0 = 1.65\mu m, \Delta\lambda = .3\mu m$), and K ($\lambda_0 = 2.20\mu m, \Delta\lambda = .4\mu m$) broadband, and $2.36\mu m$ narrowband observations.

The telescope's declination was varied to separate the telescope's constant background from the sky's background. A typical observing sequence consisted of stepping the telescope along a meridian in 20° increments.

The sun angle was varied to separate the sky's contribution from scattered sunlight within the atmosphere as well as zodiacal light.

Preliminary analysis of the K_{dark} band data indicates that the sky is significantly darker at the South Pole than at a warm site. We will publish a detailed account of the background measurement in the near future.

Our second priority was testing all of SPIREX's mechanical subsystems. This is SPIREX's first year at the pole. We have now completed 8 months of operation at the pole without any significant problems. During the 1994-1995 austral summer, we will install the necessary software to tie together SPIREX's mechanical subsystems for remote operation and automated observing.

The South Pole poses many operational hurdles. In many ways, it presents challenges that are intermediate between running a warm observatory and running a space based observatory. Issues confronted included supplying GRIM with liquid nitrogen, dealing with potential ice accumulations, and remote debugging.

Many operational issues have not proven to be the obstacles we expected. The winter-over scientist has, if anything, been spending more time outside than we anticipated. The "baby buggy cover" has made outdoor tasks much more tractable.[3] The "baby buggy cover" can be pulled over the instrument by two people. It provides a much appreciated wind brake.

In order to successfully run an observatory at the South Pole, one must be able to collaborate effectively with scientists there. The internet has proven invaluable in this regard. As more experiments are deployed to the pole, by SPIREX and other groups, it will be imperative to increase internet access and data rates to the South Pole.

During Comet Levy-Shoemaker 9's impact on Jupiter, the internet link was put to a hard test as we transferred all of the data back to Chicago in nearly real time. This amounted to roughly 25 megabytes per day from the South Pole to Chicago. Without these high data rates, SPIREX would not have been able to provide timely pictures of the comet impacts.

We have also used the internet extensively for less time critical purposes. In particular, we have a flatbed color scanner that has proven invaluable for sending scientists at the South Pole scientific papers, lab photographs, and even color newspaper articles.

We use the UNIX "talk" command to "talk" to scientists at the South Pole on a nearly daily basis. This provides more direct communication than eMail. This is important for discussing scientific issues. It also provides an important morale boost given the isolation at the South Pole.

---

[3] The "baby buggy cover" is the folding tent shown in figure 1.

In addition to background measurements, SPIREX is conducting two scientific investigations this winter. These are: (1) beginning a prototype deep survey to gain experience with the issues that will be faced in the much larger, automated, $60cm$ and $2.5m$ surveys (Hereld 1994) and (2) observing the collision of Comet Shoemaker-Levy 9 with Jupiter.

The prototype survey's goal is to begin taking deep, background limited K_{dark} band exposures. This will allow us to begin developing software to handle the data streams from automated surveys in future years.

SPIREX had a nearly uninterrupted view of Jupiter throughout the period when Comet Shoemaker-Levy 9 (figure 5) was colliding with Jupiter. Of the 20 fragments for which timing information was available, we captured 16, and saw obvious evidence of an impact in 10.

We conducted two programs during the Jupiter-comet event. Near the impact times, we took many $2.36\mu m$ narrowband exposures, and occasionally $2.22\mu m$ exposures, to track the plume's rapid evolution. In addition, we have baseline coverage between impacts at $2.22\mu m$, K band, $2.166\mu m$, $2.122\mu m$, $1.99\mu m$, and $1.54\mu m$. The baseline coverage is nominally at hourly intervals.

We are working to construct light curves of the plumes and an animated sequence of Jupiter covering much of the impact week.

5. Accomplishments

SPIREX's accomplishments this year include: (1) successfully deploying and operating a near infrared observatory at the South Pole, (2) measuring the near infrared background at the South Pole, and (3) observing Comet Shoemaker-Levy 9's collision with Jupiter.

6. 1995 Goals

Our goals for the 1994-1995 austral summer and 1995 austral winter are: (1) to integrate SPIREX's subsystems for automated remote observing and (2) to begin the $60cm$ survey.

Although all the necessary hardware for remote operation is in place, the necessary software for remote operation and automated observing are not. We will begin installing the software for remote automated observing in January 1995.

7. Summary and Conclusions

We have deployed a $60cm$ telescope, near infrared spectrometer and imager, and support systems to the South Pole as the first phase of the SPIREX project. All components of the system have functioned at the pole for seven months without significant problems.

Since April 1994, we have been monitoring the near infrared background in the K_{dark} band. The sky is significantly darker than at a warm site.

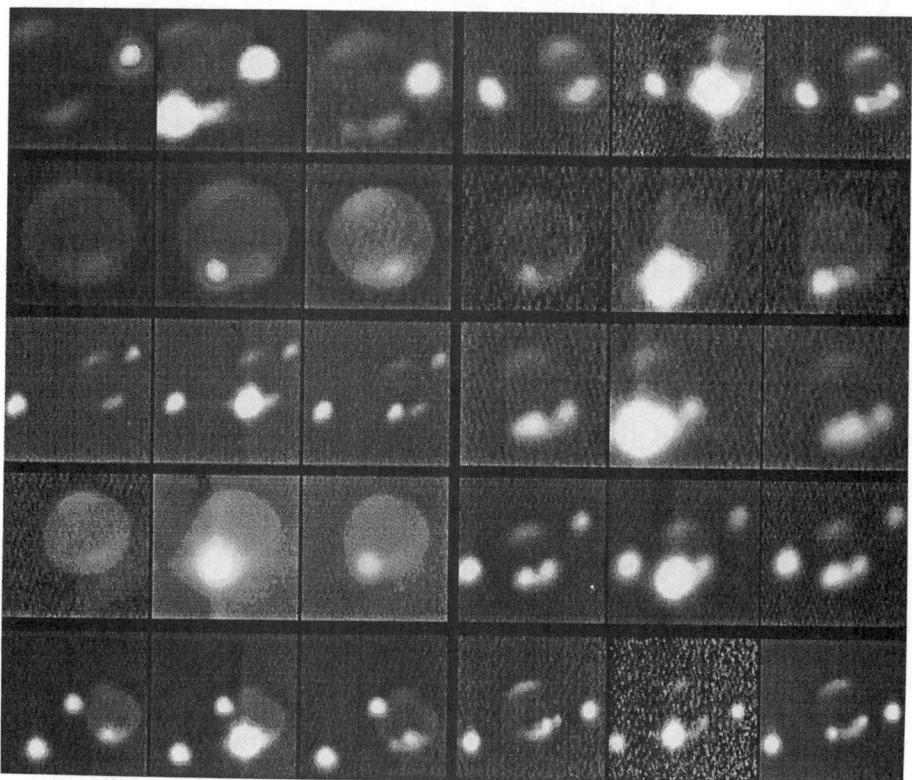

Figure 5. Comet Shoemaker-Levy 9 collides with Jupiter. These 2.36μm images are in groups of three. From top to bottom, the left hand column contains impacts A, C, E, G, and H. From top to bottom, the right hand column contains impacts K, L, Q1, R, and W. In each impact series, the first picture is Jupiter a few minutes before the plume became visible, the second picture is the plume near maximum intensity, and the third picture is the plume a few minutes after maximum intensity. These are preliminary reductions and artifacts are visible.

SPIREX observed 16 impacts from Comet Shoemaker-Levy 9's collision with Jupiter. We are working to produce 2.36μm light curves and an animated movie covering the full impact period from these data.

The internet link to the South Pole has been invaluable. As more experiments are installed at the pole, it is important to upgrade internet capability.

During the 1994-1995 austral summer we will install software necessary for automated, remote operation.

Acknowledgments. SPIREX was designed and built at The University of Chicago with the assistance of Yerkes Technical Center engineers Fred Mrozek, Dale Sandford, and Jeff Sundwall; and electronics technician Mark Thoma. D. A. Harper contributed to SPIREX's near infrared camera. Fred Mrozek was principally responsible for SPIREX's telescope.

During the six day period when fragments of Comet Shoemaker-Levy 9 were colliding with Jupiter, Yerkes Technical Center engineer John Briggs and Joe Spang (University of Wisconsin) assisted in observing at the South Pole.

1994 winter-over scientist James Lloyd assisted in the Comet Shoemaker-Levy 9 observations and data reduction in Chicago.

Like any large project, SPIREX benefited from the efforts of many people. Among these, Bob Pernic and Joe Rottman were responsible for all polar operations, without which, SPIREX would have been impossible. Bob Loewenstein designed GRIM's graphical user interface, and contributed computer expertise both at the South Pole and at home. Liz Moy-Briggs coordinated SPIREX's often complicated shipping and purchasing. The Center for Astrophysical Research in Antarctica (CARA) and Antarctic Support Associates (ASA) assisted both at home and abroad.

This research was supported in part by the National Science Foundation under a cooperative agreement with the Center for Astrophysical Research in Antarctica (CARA), grant number NSF DPP 89-20223. CARA is a National Science Foundation Science and Technology Center. Foundation Science and Technology Center.

References

Hereld, M. 1994, in Infrared Astronomy with Arrays: The Next Generation, Ian S. McLean, Dordrecht: Kluwer Academic Publishers, 248

Loewenstein, R. F., Owen, R., Yanny, B., Fowler, J. R., Harper, D. A., and York, D. 1992, Design Fundamentals of Interactive Remote Observing - Remote Usage of the Apache Point Telescope, Observing at a Distance, pb. World Scientific, pp. 19-29

Antarctic Muon and Neutrino Detector AMANDA: First Data and Outlook

J. Lynch, for the AMANDA-collaboration*

National Science Foundation

Abstract. We report the first results of the AMANDA detector (4 strings, 80 optical modules) deployed in South Pole ice at a depth of 800-1000 meters during the Antarctic summer 1993-94. We found that:

- the technology of deploying optical modules in ice is understood and surprisingly friendly. Ice is a sterile medium. The background noise in the in-situ OMs is determined by dark noise and measured to be only 1850 Hz, a factor 10-100 lower than in ocean water. The electronics and data acquisition system perform as expected,

- the absorption length of light in in-situ polar ice is 59±3 m, 2.5 times larger than that of the clearest ice studied in the laboratory and similar to that of the purified water used in the IMB and KAMIOKANDE proton-decay experiments,

- our data on scattering of light on residual bubbles are consistent with core measurements at Vostok(Byrd) where the ice is bubble-free below 1280(1100) meters,

- the data acquisition system is fully operational. First-pass reconstruction of muon tracks yields trigger rates and an angular distribution consistent with those expected for cosmic ray muons,

- redesign of the 200 optical module AMANDA, based on the experience of operating the first 4 AMANDA strings, results in an effective volume increased by a factor 5 for detecting TeV-energy neutrinos. We will position 6 additional strings at 1500 meters depth in the form of a pyramid with the existing 4 strings in the apex. The critical up-down discrimination is improved by a factor 10-100 (depending on trigger) as a result of i) deeper deployment and ii) larger spacing of the optical modules made possible by the increased absorption length. The prospects for building a kilometer scale detector in polar ice have been qualitatively improved.

1. Introduction

The high energy neutrino detector AMANDA is a multi-purpose instrument. Its science reach touches astronomy, astrophysics and particle physics. Because high energy photons, whatever their wavelength, are absorbed by a few hundred grams of matter, AMANDA's high energy neutrinos will provide us with a first opportunity to do a tomographic study of the Universe. Historic forays into new wavelength regions invariably result in the discovery of unanticipated phenomena. AMANDA will also search for the sources of the highest energy cosmic rays, dark matter, neutrino mass, monopoles and supernovae. It will be an ideal tool to pursue the exciting science pioneered by space-based gamma ray detectors such as the study of gamma ray bursts and the high energy emission from quasars.

There are compelling reasons for deploying the detector in natural ice:

i) There is no need to deploy electronics. The optical modules (OMs), sensing the Cherenkov light emitted by muons from neutrino interactions, consist of a 8 inch EMI 9353/9351 photomultiplier tube (PMT) and nothing else. The time and amplitude of the signal are carried over a cable to the electronics positioned at the surface above the detector. The high voltage is brought down to the OMs on the same cable.

ii) Ice is a sterile medium. The background noise measured in the in-situ OMs is determined by dark noise and measured to be only 1850 Hz, a factor 10-100 lower than in ocean water.

iii) It was thought that the relative short attenuation length of 25 meters, established in laboratory measurements, was somewhat of a drawback for deploying Cherenkov detectors in polar ice. After operation of the first 4 AMANDA strings we found however that natural ice is very clear with an absorption length of 60 meters.

In this note we discuss the deployment and first results of the 4-string AMANDA detector shown in Fig. 1. We discuss the prospects of using ice as a particle detector and present a Monte Carlo study of an optimized detector architecture constructed with an additional 78 OMs. We also speculate on the design of a 1 kilometer scale detector.

2. Technology

Contrary to drilling experience gained from the 91-92 deployment of 2 small strings, deploying AMANDA strings using the new drills was surprisingly easy. These drills were designed to operate in cold South Pole ice unlike the 91-92 drill which had operated in the Arctic where the ice is 20°C warmer. The winches are now computer assisted. The hot water drilling technique has been firmly established. The bottom line is that

- 4 holes were drilled 1 km deep, the diameter was 60 cm,
- water at 90°C was pumped down the holes at a rate of 40 gallons per minute,
- the time to drill a hole was less than 90 hours, typically 4 days,
- fuel usage was about 3000 gallons per hole.

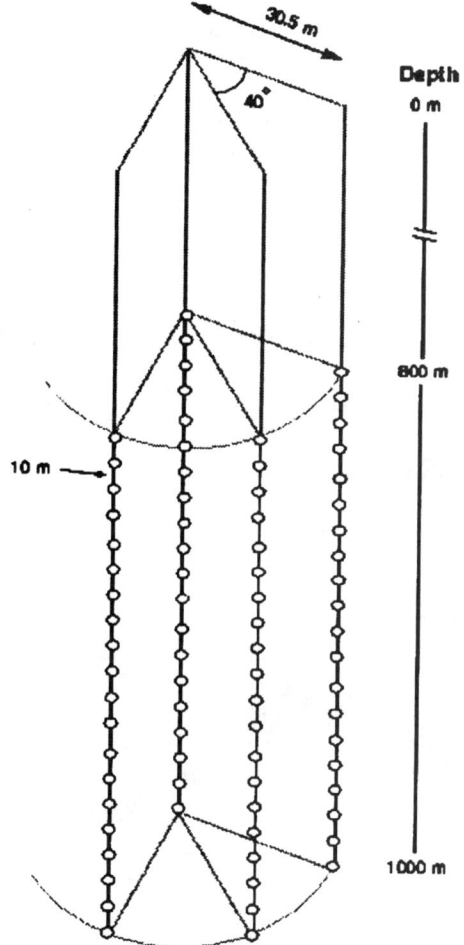

Figure 1. Perspective view of the 80 deployed AMANDA optical modules. Each of the 4 strings contains 20 optical modules separated by 20 meters.

Four strings with 20 OMs each have been positioned at depths between 800 and 1000 meters. We lost 3 out of 80 OMs during deployment. Four other OMs, although operating, are to a varying degree problematic. Most of these problems were associated with the first string. We adjusted our deployment procedures and, except for a single OM, strings 3 and 4 are perfect. Operation of the detector has been totally stable since January 94.

The background noise in the OMs, measured in-situ, is only 1850 Hz; see Fig. 2. This is below the 4 KHz observed in the prototype(Lowder 1993) positioned at shallow depth. The OMs are operated at the single photoelectron level. The peak-to-valley of in-situ photomultiplier tubes is shown in Fig. 3.

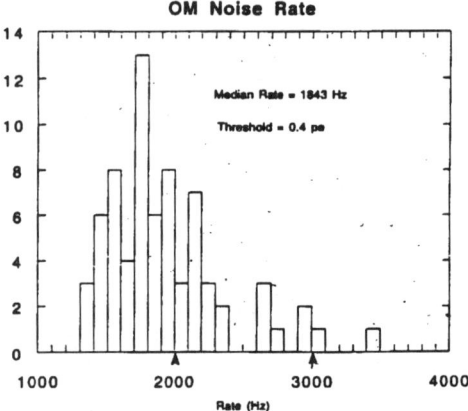

Figure 2. Distribution of in-situ noise rates of deployed AMANDA photomultiplier tubes.

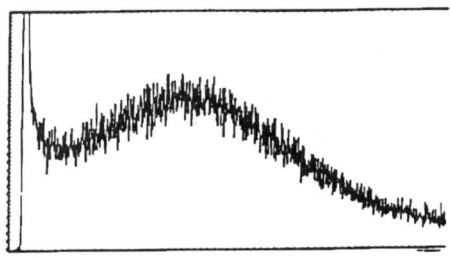

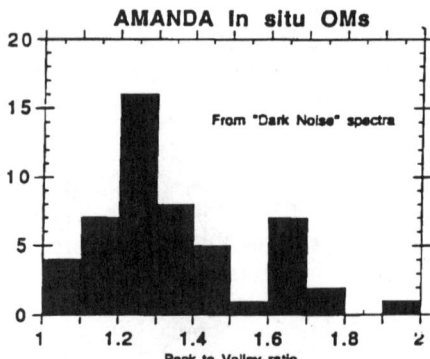

Figure 3. Single photoelectron peak-to-valley in AMANDA photomultiplier tube and peak-to-valley distribution of in-situ photomultiplier tubes.

Also deployed was a laser calibration system which pulses light into a nylon diffuser ball positioned 30 centimeters below each OM. This system is fully functional though the light intensity is degraded over some of the fibre optic cable. This is not a handicap as the system is extensively overconstrained.

The data acquisition system is fully operational. Before filtering, 40 MBytes of data are collected every hour at a trigger rate of 22 Hz. The data sent back in real time via satellite is 10 MBytes per day now and 50 MBytes in the future. Calibration data has also been taken and mapping of the detector is complete though improvements are made as data analysis progresses.

3. Properties of Ice

AMANDA was designed on the information that(Gow and Williamson 1975)

- the maximum attenuation length of light in enclathrated, bubble-free ice is 25 meters based on laboratory measurements,
- ice becomes enclathrated at 400 meters depth in −55°C ice,
- any remaining bubbles (none were predicted) would have a radius of less than 1 micron at 1 km depth.

The AMANDA group itself established a minimum attenuation length of 20 meters from in-situ measurements in Greenland(Lowder et al. 1991) and at the South Pole(Lowder 1993) and from coincidences of the SPASE array with a pair of 3 inch photomultiplier tubes positioned at a depth of 800 meters. The latter measurement indicated that muons in SPASE-triggered showers were collected from further distances than anticipated on the basis of a 25 meters attenuation length of the light(Tilav et al. 1993). The collection area of the 3 inch PMTs was too small to establish the effect with confidence.

With 4 AMANDA strings as well as a laser calibration system in place we now have the potential to gather information on in-situ ice with experimental methods superior to those available to glaciologists. Detailed measurements of in-situ ice have been completed(AMANDA coll.) exploiting the laser calibration system. A pulsed nitrogen laser is used to drive a dye laser which pulses 500 nm light with a frequency of 20 Hz into the fibre optic calibration system. The nylon spheres radiate light isotropically which is detected by neighbouring PMTs and by PMTs on neighbouring strings. The time resolution of each PMT is 2 ns. By measuring the relative timing of the signals, both the optical properties of the ice and the position of the tubes can be accurately derived. The timing data are well described by a random walk model which includes the scattering of the light on residual bubbles. A sample of timing data, compared to this model, is shown in Fig. 4. The measurements can be repeated at different depths in the detector. Our conclusions are that:

- Ice is not completely enclathrated at 1000 m. The scattering length of the light on residual bubbles is shown in Fig. 5a. It decreases linearly with depth and, assuming a linear extrapolation, vanishes at 1100 meters. Our data on the scattering length is consistent with the one inferred from ice cores of similar temperature at Vostok(Badd et al. 1984) where residual bubbles with a radius of 50 microns, approximately independent of depth, and a density of $200/cm^3$ imply a scattering length of order 1 meter at a depth of 800 meters. In Fig. 5b we compare our results with scattering lengths inferred from cores retrieved at Vostok and Byrd station where ice becomes bubble-free at depths of 1280 and 1100 meters, respectively.

- The absorption length of the light is 59 ± 3 meters, a factor 2.5 larger than the value obtained in previous measurements: see Fig. 6.

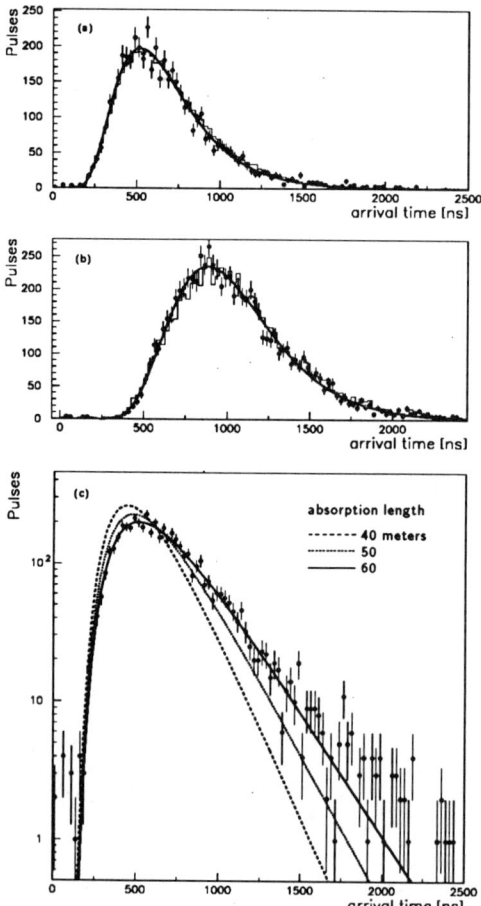

Figure 4. Time distribution of laser light as detected by the optical module after travelling 21 (a) and 32 (b) meters. The data on light propagation is reproduced by a Monte Carlo which includes absorption and scattering on residual bubbles. Sensitivity to the absorption length is illustrated in (c) where the data in (a) is fitted assuming an absorption length 40, 50 and 60 meters.

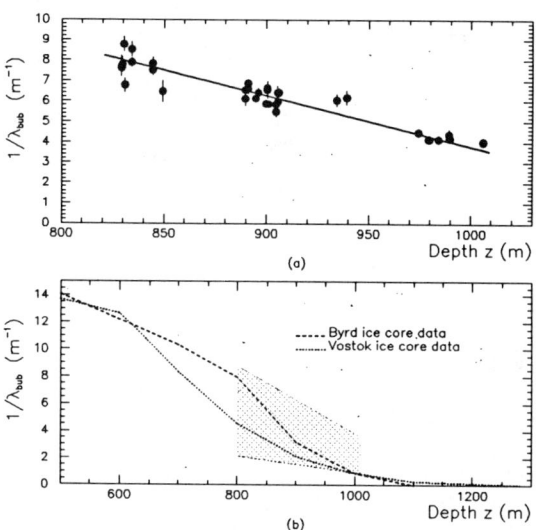

Figure 5. (a) The inverse scattering length on residual bubbles as a function of depth. The slope is 0.08 cm/meter in the interval between 800 and 1000 meters. By linear extrapolation the inverse scattering length vanishes at 1100 meters depth. (b) Comparison of the bubble density inferred from the laser data(AMANDA coll.) with the density measured in cores at Vostok and Byrd station.

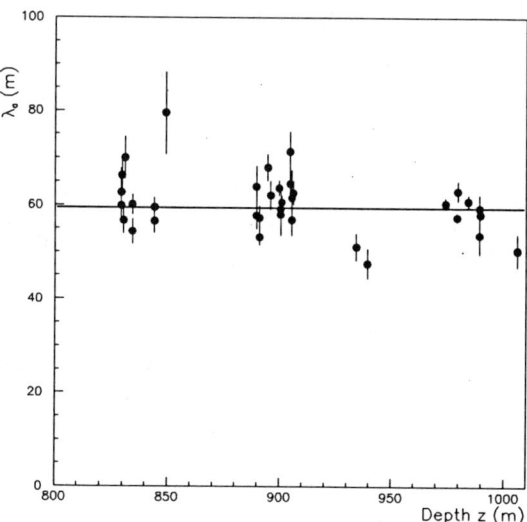

Figure 6. Absorption length deduced from the timing distributions of pulsed laser light as a function of depth.

Our conclusions are qualitatively supported by a study of the time structure of the light associated with cosmic ray muons sweeping through the planes of 4 PMTs at 20 consecutive depth levels. Also the trigger rates of muons observed in coincidence with the SPASE air shower array provide information on the properties of the ice. This work is in progress.

4. Status of muon reconstruction

Light collection of the photomultiplier in bubbly ice is more efficient for muons with impact parameter less than 10 meters. Beyond that distance the likelihood of triggering drops sharply. The reason is that beyond 10 meters the actual distance travelled by the photon in the random walk becomes large compared to the 60 meter absorption length. Notice that one can collect photons over more than 10 meters if the source is sufficiently bright as is the case for the laser pulses. The relative count at 30 meters varies by more than an order of magnitude for a change in absorption length from 25 to 60 meters and the absorption length can be determined with high precision; see Fig. 4(c).

Reconstruction of muons in the presence of large angle scattering on bubbles is rather straightforward provided the propagation of light in the detector medium is adequately understood. This information is provided by the laser calibration system. The detector basically operates as a drift chamber rather than a Cherenkov detector and the track fitting procedure should be adjusted accordingly. A muon traversing the instrumented ice is not accompanied by the familiar Cherenkov bow wave. The light pattern is that of a an isotropic light source travelling through the ice with speed c/n. Using the timing information obtained with the laser calibration system, we can determine, for each OM in the trigger, the expected number of photons and their arrival times as a function of the module's impact parameter relative to the muon track. Also the expected spread in arrival time is known. This information is used to determine the muon direction. If the information is inadequate to reconstruct the track direction with sufficient precision, the event is rejected, thus reducing the effective area of the detector.

With only a multiplicity (larger than 8) and a time-over-threshold cut 90% of muons can be reconstructed with sufficient precision to obtain trigger rates and a zenith angle distribution consistent with those expected for comic ray muon rates; see Fig. 7.

Muon reconstruction is possible provided that the radiation pattern of the muon is precisely determined. Work is in progress designing quality cuts which will determine the ultimate angular resolution that can be achieved with the present 4 strings.

5. SPASE-AMANDA coincident triggers

Coincidences with the SPASE surface air shower array have been observed. These events are now used as a calibration tool. The large rate of coincidences (more than 100 per day for a 10-fold trigger), significantly higher than predicted, gave us a first hint that the strings detected muons at larger distances than expected. A sample of coincident muon events is shown in Fig. 8. The background

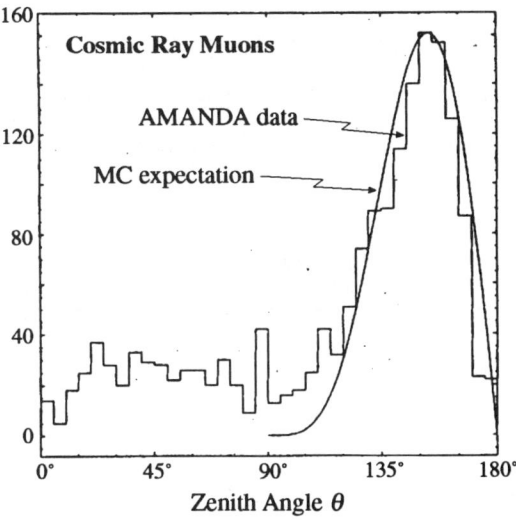

Figure 7. First-pass reconstruction of cosmic ray muon triggers is compared with the Monte Carlo expectation.

is consistent with random cosmic ray muons triggering the detector within the SPASE-AMANDA coincidence time window. The 4-string AMANDA is by a factor the largest muon detector ever built. Its relative area compared to the MIA detector in Utah or the MACRO, LVD detectors in the Gran Sasso tunnel depends on the composition of the cosmic ray primaries. The data can be used to study the response of ice to a triggered muon beam. As the direction of the muon is known the data can be exploited to finetune the procedures we have developed for reconstructing muon direction in the presence of residual bubbles.

The AMANDA detector is the first muon telescope large enough to detect muons, with energy in excess of 180 GeV, without sampling. The complete muon bundle contained in SPASE showers is observed. Although AMANDA does not directly measure muon-energy, the coincidence rate can be studied as a function of trigger multiplicity. The data are being studied with the goal to make a qualitatively improved determination of the elemental cosmic ray composition in the region of the "knee" in the primary spectrum. Heavy nuclei of high energy produce more muons than protons, hence the sensitivity to composition of the primary cosmic ray beam. It is well-known that this represents one of the most fundamental measurements in cosmic ray astrophysics.

6. Bubble-free Ice as a Particle Detector

We anticipate that deep polar ice is bubble-free with an attenuation length as large as the one in distilled water used in detectors such as IMB and KAMIO-KANDE. Can this fantastic clarity be exploited? The answer depends on the scattering on particles and refraction through clathrate crystals in the bubble-free ice. We discuss this next.

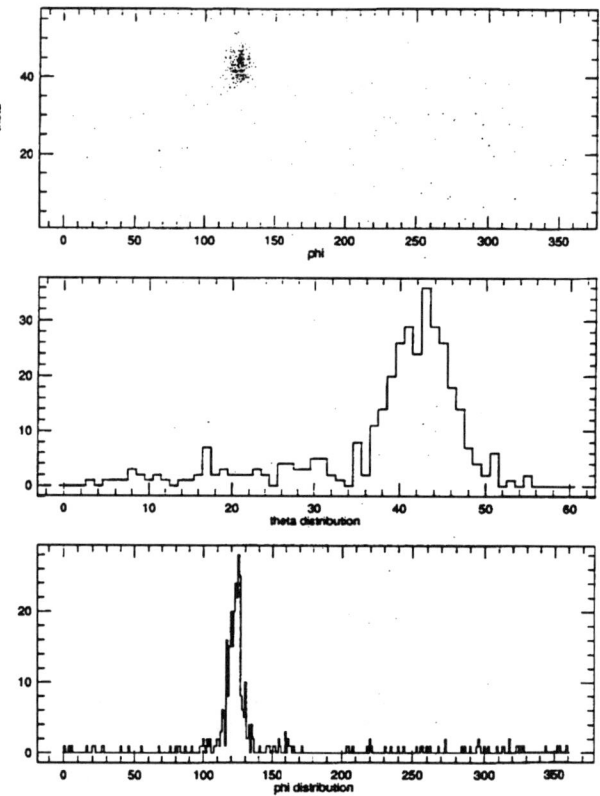

Figure 8. Sample of AMANDA muon triggers (10 or more) coincident with SPASE showers. The data is plotted in terms of the 2 polar angles defining the direction of the air showers as determined by SPASE. As viewed from the air shower array the AMANDA detector is positioned at zenith angles $\theta \simeq 45°$ and $\phi \simeq 120°$.

Absorption sets an upper limit on the spacing between phototubes. Other effects limit the quality of the "image" of the muon's trajectory: scattering by dust and by bubbles, and refraction at boundaries between ice crystals and at boundaries between the air hydrate phase and the normal hexagonal ice phase. Until now, the four effects have been studied only by microscopic examination of sections of ice cores from various locations in Antarctica and Greenland. Moseley-Thompson(Moseley-Thompson) has studied dust in a South Pole core at depths down to only 349 m, which is not nearly deep enough for our purposes.

6.1. Scattering by Dust

Paleoclimatologists have studied ice-cores in order to determine changes in temperature, dust accumulation, and air composition(Thompson 1977, Lorius et al. 1979, Petit et al. 1981, Gayley and Ram 1985, Lorius et al. 1985, Jouzel et al.

1991). The depth dependence of the isotropic composition of oxygen bubbles ($^{18}O/^{16}O$) provides a measure of temperature as a function of time in the past. Dust grains, originally present as atmospheric aerosols, serve as nucleation sites for precipitation. The dust concentration at a given depth in an ice core correlates strongly with temperature at the time of deposition: there is a causal relationship between ice ages and aerosol concentration. The concentration increases by as much as two orders of magnitude during an ice age. A study of light scattering and absorption by dust suspended in a Dome C ice core(Royer et al. 1983) is particularly useful in that they directly measured the quantities of interest to AMANDA. From their data we infer a scattering length of ~ 10 m for dust corresponding to the Last Glacial Maximum (LGM, 18,000 years ago) and a scattering length of 20 to 30 m for an interglacial layer. For our 20 cm PMTs, the latter value is probably acceptably small given that the scattering is mostly forward.

To predict age as a function of depth in the South Pole ice, it is useful to construct a universal function, shown in Fig. 9, in which reduced depth, $d^* = d/H$, is plotted as a function of reduced age, $t^* \equiv t\,a/H$, where a = accumulation rate and H = ice thickness to bedrock. Knowing a and H for the South Pole, we can estimate age vs. depth from this graph. The finite dispersion of the points results from the neglect of differences in lateral flow rate, which is zero only at an ice divide. Using this graph, we find that the dust maximum due to the LGM, seen in the Vostok core at 340 to 500 m and in the Dome C core at 500 to 700 m, probably occurs at 800 to 1250 m at the South Pole. We estimate that the next maximum in the dust concentration, corresponding to an age of ~ 60 kyr, occurs at a depth of ~ 2600 m at the South Pole.

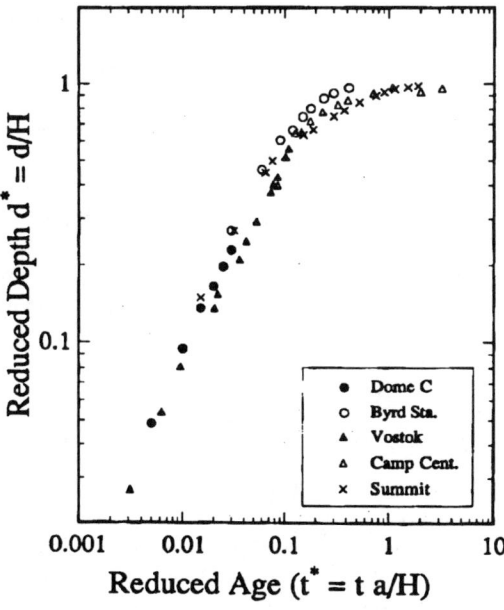

Figure 9.

6.2. The Bubble Problem

The concentration of trapped air below the firn layer depends on atmospheric pressure and thus on the elevation of the surface. The concentration and size of bubble as a function of depth depend on a number of factors including pressure, temperature, crystal size, and dust concentration. An experimental complication is that secondary bubbles may form when a core is removed from the ice. P. Duval (private communication) has remarked that due to the higher accumulation rate at the South Pole than at Vostok, the transition zone within which bubbles begin to transform into air-hydrate crystals may occur about 200 m deeper at the South Pole. More important to us is the depth for disappearance of bubbles at the South Pole. Bubbles are completely absent at depths below 1100 m at Byrd Station and below 1280 m at Vostok.

6.3. Refraction by Air Hydrate Crystals

Cherenkov light will be deviated from straight-line trajectories by refraction from air hydrate crystals. The depth at which the air becomes enclathrated instead of trapped in bubbles is predicted by MillerMiller 1969) to range form $\sim$ 1000 m for ice at $-20°C$ to $\sim$ 400 m for ice at $-50°C$. For the Dye-3 and Camp Century cores, the observed depths for air hydrate appearance(Shoji and Langway 1987) agree with Miller's calculations; for Byrd Station, the observed depth for the appearance(Shoji and Langway 1987) is the $\sim$ 100 m shallower than that calculated by Miller.

The effect of these crystals on light transmission in deep ice will depend on their size, shape, concentration, and refractive index relative to that of the normal ice in which they are imbedded. Langway and his collaborators have made X-ray and optical studies of air hydrate crystals in ice cores(Shoji and Langway 1987), (Shoji and Langway 1982), (Hondoh et al. 1990). T. Uchida(Uchida) has measured a mean value of 1.004 for the ratio of refractive indices of clathrate hydrate crystals and of hexagonal ice in a Vostok core. He and his collaborators(Uchida et al. 1994) find a typical concentration of $\sim$ 500 hydrate crystal per cm^3, a typical diameter of $\sim$ 100 microns, and a predominance of spherical shapes, from which we infer a typical mean path of $\sim$ 25 cm for encounter of a light ray with such crystals. To reach a phototube at a distance of, say, 25 m requires $\sim 10^2$ traversals of such crystals. For a typical angle of incidence of $45°$, a refraction at both entrance and exit, and 100 encounters, the net lateral deflection due to a random walk about the initial direction would be less than 2 cm over the 25 m pathlength. Thus, air hydrate crystals have virtually no effect on the imaging of muon trajectories.

6.4. Refraction at Ice Crystal Boundaries

Ice, with a hexagonal crystal structure, has a refractive index along the c-axis that is larger than that in the basal plane by a factor of 1.001. Assuming a mean crystal size of 4 mm, the same as measured in the Vostok core at a depth of 1600 m, a light ray in South Pole ice would refract $\sim$ 6000 times in traversing a distance of 25 m. Although it is likely that, as is the case for cores from Dome C, Byrd, Camp Century, and Dye-3, the c-axes are strongly aligned along the vertical direction, we cannot rule out the possibility that, as is the case for Vostok and Mizuho cores, the c-axes show a more complicated distribution.

To overestimate the effects, we assume a random distribution of c-axes with a typical angle of 45° between c-axes in adjacent crystals. The net deflection due to a random walk of 6000 steps would be only 0.03 cm over the 25 m pathlength!

We conclude that, at the depth where all bubbles have transformed into air hydrate crystals and where the concentration of insoluble particles is characteristic of on interglacial period, the South Pole ice will be an adequately clear medium, and arrival times and trajectories of Cherenkov photons will be virtually unaffected by hydrate crystals and by birefringence of the ice.

What next?

One of the logistics advantages of building a detector in polar ice is that deployment of OMs is not restricted to a rigid predesigned frame. The detector architecture can be reconfigured after each deployment phase and therefore design can follow physics. The message from this year's data is clear. Further OMs should be deployed deeper, in order to avoid residual bubbles, and implement larger spacings between strings as well as between OMs on a string in order to exploit the larger absorption length. Our preliminary design for the next phase of development, shown in Fig. 10, adds six strings to form a pyramid in which last year's 4 strings form the apex and the base consists of a pentagon with one string at its center. The 5 strings will be deployed on a circle of 60 meter radius, preferably at a depth of 1500 meters. Each string contains 13 OMs with a spacing of 20 meters. The pentagon contains only 65 OMs. Its effective volume of 2.5×10^6 m^3, by itself, is already larger than the previous 200 OM AMANDA design(Tilav et al. 1993). Volumes quoted are for 7/3 triggers. Adding a deep string at the center will raise the volume to nearly 10^7 m^3. This is almost an order of magnitude larger than the original AMANDA volume. The threshold of the detector has, unfortunately, been increased. In the original design detection efficiency rose to full efficiency between 2 and 20 GeV. The threshold curve is shifted by a factor 5 for the new OM-architecture.

The up-down rejection should be significantly improved with i) deeper depth and, therefore, reduced cosmic ray muon background and ii) the larger spacing of the OMs which implies a larger sampling of the muon tracks. Monte Carlo simulations indicate a factor 10-100 improvement depending on the trigger multiplicity.

The positive implications of our findings for the future deployment of a 1 km-scale detector between 1 and 2 kilometer depth are self-evident. For the construction of a 1 km scale detector one can imagine the second phase AMANDA detector which consists of 5 strings on a 60 meter radius circle with a string at the center (referred to as a $1 + 5$ configuration) as the basic building block with an effective volume of 10^7 m^3. Imagine AMANDA "supermodules" obtained by extending the string length (and module number per string) by a factor close to 4; see Fig. 11. Supermodules would then consist of $1 + 5$ strings with, on each string, 51 OMs separated by 20 meters for a length of 1 km. A 1 km scale detector then might consist of a $1 + 7 + 7$ configuration of supermodules, with the 7 supermodules distributed on a circle of radius 250 m and 7 more on a circle of 500 m. The full detector contains 4590 phototubes which is

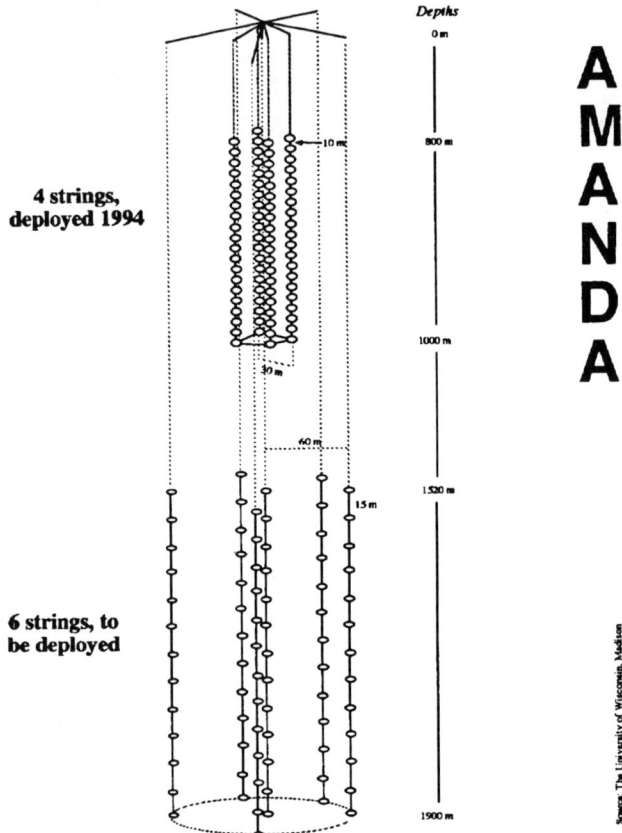

Figure 10. AMANDA, redesigned.

less than the 7000 used in the SNO detector. Such a detector can be operated in a dual mode:

- it obviously consists of roughly 4×15 the presently designed AMANDA array, leading to an effective volume of $\sim 6 \times 10^8$ m^3. Importantly, the characteristics of the detector, including low threshold in the GeV-energy range, are the same as those of the AMANDA array module.

- the 1+7+7 supermodule configuration, looked at as a whole, instruments a 1 km^3 cylinder with diameter and height of 1000 m with optical modules. High-energy muons will be superbly reconstructed as they can produce triggers in 2 or more of the modules spaced by large distance. Reaching more than one supermodule requires 50 GeV energy to cross 250 m. We note that this is the energy for which a neutrino telescope has optimal sensitivity to a typical E^{-2} source (background falls with threshold energy, and until about 1 TeV little signal is lost).

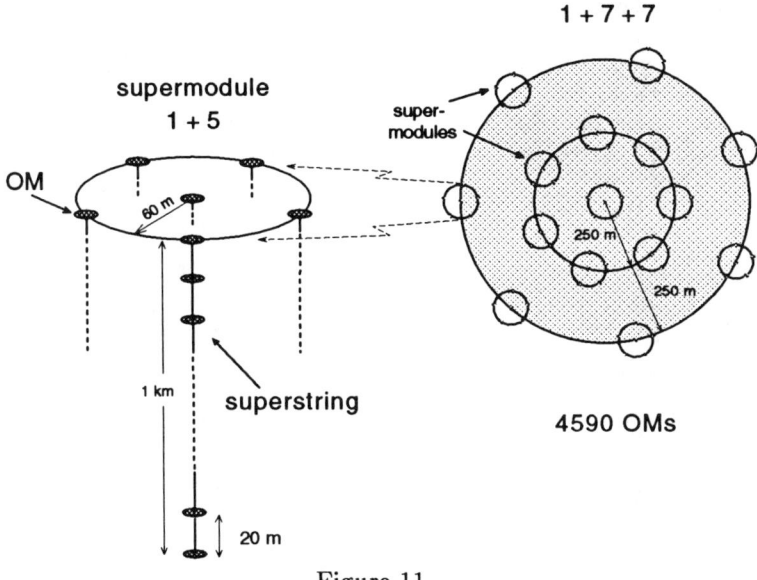

Figure 11.

What are the estimated construction costs for such a detector? AMANDA's strings (with 10 OMs) cost $150,000 including deployment. By naive scaling the final cost of the postulated 1 + 7 + 7 array of supermodules is of order $50 million and still below that of Superkamiokande (with 11,200 × 20 inch photomultiplier tubes in a 40 m diameter by 40 m high stainless steel tank in a deep mine). It is clear that the naive estimate makes several approximations over- and underestimating the actual cost.

*Members of the AMANDA Collaboration are:
P. Askejber[a], S. Barwick[b], L. Bergström[a], A. Bouchta[a], S. Carius[c], A. Coulthard[d], K. Engel[d], B. Erlandsson[a], A. Goobar[a,k], L. Gray[d], A. Hallgren[c], F. Halzen[d], P.O. Hulth[a], J. Jacobsen[d], S. Johansson[a,e], V. Kandhadai[d], I. Liubarsky[a], D. Lowder[f], T. Miller[f,g], P. Mock[b], R. Morse[d], R. Porrata[b], P.B. Price[f], A. Richards[f], H. Rubinstein[c], J.C. Spang[d], Q. Sun[a], S. Tilav[d], C. Walck[a] and G. Yodh[b].

[a]Stockholm University, Sweden
[b]University of California, Irvine, CA
[c]Uppsala University, Sweden
[d]University of Wisconsin, Madison, WI
[e]Currently at Jönköping University, Sweden
[f]University of California, Berkeley, CA
[g]Currently at Bartol Research Institute, DE

Acknowledgments

This work was supported by the National Science Foundation, USA, the K.A. Wallenberg Foundation, Sweden, The Swedish Natural Science Research Council, The G. Gustafsson Foundation, Sweden, Swedish Polar Research, Sweden, and the Graduate School of the University of Wisconsin, Madison, USA.

References

Lowder, D. M. et al. 1993, 23rd International Cosmic Ray Conference, Calgary, Canada

Gow, A. J. and Williamson, T. 1975, Cold Regions Research and Engineering Lab. Research Report 339 and 1991, private communication; see also reference 10

Lowder, D. M. et al. 1991, Nature, 353, 331

Tilav, S. et al. 1993, 23rd International Cosmic Ray Conference, Calgary, Canada

AMANDA collaboration, Nature (submitted for publication).

Badd, Yu. F. et al. 1984, Moscow Glaciological Research Publication.

Moseley-Thompson, E., private communication.

Thompson, L. G. 1977, IAHS Publication, 118, 351
  Lorius, C. et al. 1979, Nature, 280, 644
  Petit, J.-R., Briat, M. and Royer, A. 1981, Nature, 293, 391
  Gayley, R. I. and Ram, M. 1985, J. Geophys. Res., 90, 921
  Lorius, C. et al.1985, Nature, 316, 591
  Jouzel, J. et al. 1993, Nature, 364, 407

Royer, A., DeAngelis, M. and Petit, J.R. 1983, Climatic Change, 5, 381

Miller, S. L. 1969, Science, 165, 489

Shoji, H. and Langway Jr., C.C. 1987, J. de Physique, Coll. C1, suppl. to no. 3, vol. 48, p. 551

Shoji, H. and Langway, Jr., C.C. 1982, Nature, 298, 548

Hondoh, T. et al. 1990, J. Inclusion Phenomena, 8, 17

Uchida, T. private communication.

Uchida, T., Hondoh, T., Mae, S., Lipenkov, V. Ya. and Duval, P. 1994, J. Glaciol., 40, 79

PART III

FUTURE PROSPECTS

Small Telescopes and NAO's New Venture

David L. Crawford
KPNO/NOAO[1], P.O. Box 26732, Tucson AZ 85726 U.S.A.
Email: dcrawford@noao.edu

Abstract. Small telescopes have a great deal of value in astronomical research and in science education, yet their cost is relatively low compared to large telescopes. As such, they may be the only effective way for most astronomers to get an adequate number of observing hours for their programs. Recognizing this fact, as well as their own need for small telescopes (here defined as 2.5m and smaller), NOAO has embarked on a new program, within its technology transfer mission, to partner with one or more companies producing small telescopes to develop a new series of such telescopes and to market them widely to interested organizations worldwide. Anyone interested, for any reason, is encouraged to contact NOAO's Tech Transfer Office.

1. Introduction

There is no doubt that relatively small telescopes, here defined as 2.5m and smaller, have a great deal of value in astronomy, both for research and for education. With the modern detectors and instrumentation, a one meter telescope is as powerful for imaging and wide band photometry as was the 200-inch telescope only a few decades ago. Furthermore, small telescopes are essential as complementary tools in support of larger telescopes, as well as space astronomy and radio astronomy. In addition, they are the only viable way to do extensive monitoring and surveys of all sorts.

Let us look at the value issue in a bit more detail. It is clear that the cost of a telescope scales with something between the 2.5 and 3.0 power of the aperture, if we compare ones of similar technologies or similar epochs. Past studies have shown that a 2.7 power is perhaps the best fit. However, the number of users or the number of papers clearly scales with a lower power. I have updated Abt's earlier study and find powers of 1.0 or less to be a good fit to existing data. It is clear therefore that the Value Per Cost ratio is excellent for small telescopes. It does not mean that we don't need large telescopes too, for each size does things better than other sizes do. I have discussed these issues in some depth in an earlier paper (Crawford, 1994) and won't repeat the arguments here except to note a few of the items where small telescopes have special value: they cost less than large telescopes, they serve about the same number of users as does

---

[1] Operated by AURA Inc under cooperative agreement with the National Science Foundation.

a large telescope, they have a good match of pixel size to seeing, they have larger fields-of-view for the same linear scale at the aperture, they make great training grounds for large telescope users, and they are essential complements for research at large telescopes and for space facilities. Most of these things are self evident, but it is worthwhile to continue to repeat them, so that we do not get a fatal case of "aperture fever." Astro-economics is a valuable science, especially in times of tight funding.

2. NOAO and Small Telescopes

In recent years, NOAO has closed a number of small telescopes, shutting off a number of users therefore. Such a scenario might well be expected to continue, in view of the upcoming eight meter telescopes and their impact on NOAO's budget. However, it is clear to management that small telescopes are an important complement to the large ones, and that new generation small telescopes are a very viable tool for modern research. In addition, NOAO has begun a modest program of technology transfer, with the view of sharing its technology resources with others. One such program involves small telescopes.

This Spring, we held a meeting with a number of companies that are active in small telescope development or production, as well as several potential customers for such telescopes. All attendees expressed strong interest in the issues. As a result, we plan to continue discussions with such companies with the goal of partnering with one or more to develop (or refine the current developments) of a set of new generation small telescopes, and then to market such telescopes worldwide. The goal is to have a "standard design," perhaps with several flavors, that would be very efficient, and with relatively low maintenance costs, and one that could be built in volume so as to help minimize costs. In a sense, one wants to maximize the Value Per Cost ratio.

We expect also to partner with a company to develop and market a new telescope control system, based on the present work done at NOAO on the WIYN and the 4m telescope control systems. NOAO alone has a market for about a dozen of such control systems to upgrade the present telescopes, and the world market for such a system should be large.

3. Summary

We think that the value of small telescopes is large, and there is no question that their cost is relatively low compared to large telescopes. Such telescopes are a viable complement to research with large telescopes and with space facilities. NOAO has the need for several such telescopes as a complement to the upcoming two eight-meter telescopes, as well as to provide telescope time on first rate facilities to its many users. Therefore, one of its technology transfer programs has the goal of partnering with one or more industrial companies to develop and market a set of new generation small (2.5m aperture or less) telescopes, as well as a control system for such telescopes. Dialogs with such companies have begun, and we hope to select one or more before mid-1995.

We urge interested parties, developers, users, anyone, to contact us with their input, suggestions, or questions. Please do stay in touch; we will plan to do so with all who contact us. As in any viable technology transfer program, we must be "customer driven." We need your input and your involvement.

References

Crawford, D. L. 1994, Paper given at Kilkenny (Ireland) Conference of Astronomical Photometry, in press

The Rationale for an Automatic Spectroscopic Telescope

Joel A. Eaton
Center of Excellence in Information Systems, Tennessee State University, Nashville, TN U.S.A.

Abstract. Computers have changed our lives by making it possible to control large amounts of information effectively. This, in turn, makes the manufactured goods we use cheaper and makes them better. An entirely parallel development is occuring in astronomy, for the same reasons. We discuss the application of computer-based automation to astronomy, specifically to observing, and discuss reasons for extending it to spectroscopy.

1. Introduction

Over the past twenty years there has been a revolution in manufacturing brought on by application of computers to planning and controlling complex operations. This change has been accompanied by a bewildering number of acronyms as well as unjustified hyperbole, but the fact remains that computers have made manufacturers a lot more efficient as well as nimbler, with marked effects on employment. In 1980, when the fraction of Americans working in manufacturing was near 20%, a delphic study predicted that such employment would dip to 12% by the end of the century. Alarmist stories in our newspapers over the past decade seem to bear out this predicted trend.

One advantage of computers in making things is the way they can control actual machines very precisely, so much so that manus no longer has an advantage. Machines do their jobs uniformly so that a constant level of quality can be maintained. Computers are rather single-minded, ready to send the right commands to a machine to do just what a design engineer had in mind, or at least what he told the computer to do.

The greater influence of automation has been on the design engineer's own job, however. Computers are now being used to design and test products electronically. Data are transferred from department to department much faster, eliminating delay, duplication, and the inevitable associated errors. Designs may now be tested in computers and parts that would interfere modified, thus eliminating full stages of costly mockups. New designs are being done as modifications of existing designs, with great savings in unnecessary duplication. Designs are being simplified for more efficient manufacturing, thus improving their reliability. Finally, the speed that automation has allowed gives manufacturing the increased nimbleness to produce small lots of goods efficiently and to begin making customized products in large factories.

Figure 1. The future of observational astronomy. Shown is the APT farm managed by Lou Boyd atop Mt Hopkins in Arizona. Each of these telescopes observes automatically, under computer control, usually with nobody on site. TSU manages four of the present eight telescopes in this facility. Others are managed by principal astronomers at Villanova, Franklin and Marshall, the College of Charleston, and Arizona State.

The result of all this is that employment is dropping not because industry is failing but because it is so much better that it needs fewer workers. Its products have become cheaper and better.

2. Automation in Astronomy

At Tennessee State University we are convinced that automation is the future of observational astronomy. We have gained this perspective from our experience with automatic photometric telescopes (see Fig. 1) and with using computers to plan observing and analysis of observational data. The traditional arguments in favor of manual observing–that it's romantic to go to the observatory and that manual observing gives better control of results and higher quality–lose their vigor in the cold light of experience. Ignoring the romance of Luddism, we find automation has made our photometric observing both cheaper and a lot more effective. Thus the reasons for automating astronomical observing are exactly the same as for manufacturing. We break down these advantages as follows:

1. **Increased Efficiency:** Data can be gathered more economically by automatic robots than by manually controlled machines.

    a. <u>Individual observations are less expensive.</u> Although allocation of expenses to particular projects is subject to criticism, the cost per observation does seem considerably less with robotic telescopes. Henry (see "The Fairborn/TSU Robotic Telescopes Operations Model," this volume) has compared costs of observing trips he and I made to Kitt Peak in the early 1980's with routine observing with APT's. He finds the manual observing was about twenty times as expensive per differential photometric observation of a variable star. These dramatic savings were realized in time not spent at the telescope as well as in the large amounts of time spent reducing data that had to be manually corrected for the inadequate data logging system used at Kitt Peak.
    b. <u>More data can be gathered per star.</u> Automatic telescopes can observe during all clear weather available. Manual observers rarely do, even if they have the stamina to try. Thus with robotic telescopes analyses are less likely to be based on an inadequate number of data.
    c. <u>Astronomer's time is used more effectively;</u> in planning observations and analyzing data. Given current salaries and overhead rates, an astronomer's time is worth about $50 per hour. This is too valuable a quantity to waste on such repetitive tasks as manual observing.

2. **Increased Quality:** The same qualities of automation that make it efficient also make it effective.

    a. <u>All data are gathered in the same way.</u> This is especially important, since it means that quality can be defined and causes for degradation identified. Quality is thereby improved incrementally with elimination of a succession of problems (see Henry, "The Development of Precision Robotic Photometry," this volume).
    b. <u>Planning for quality is easier and indeed necessary.</u> Automated observing places much more rigorous demands on equipment than does manual observing. Automated instruments must be strong enough to stand up to mind-numbing repetition. There will be no human observer to fix them when they break or to compensate for their maladjustments. Such machines must therefore be designed so they won't break in the many ways that manual observers have traditionally been expected to tolerate. When robotic telescopes do break, they must be easy to diagnose and to fix, as well. Furthermore, an automated observing program must be designed with the same attention to quality. It must incorporate a strategy for obtaining routine calibration data and quality-control observations. These ancillary data then form a database with which the stability of the instrument can be assessed and incremental improvements planned. Such planning is very difficult in the typical Balkanised manual observing at communal facilities.

c. <u>Different kinds of observing programs can be attempted.</u> Traditional observing has been based on short observing runs lasting from several nights to several weeks. Most professional astronomers cannot afford the living expenses to use more than a few of these sessions each year. The result is that projects requiring long strings of data cannot be done effectively, and astronomers are effectively limited to projects that do not require long-term monitoring or amassing huge numbers of observations. Automation is not subject to such constraints, and this makes a whole new group of studies possible.

3. Automating Spectroscopy

The astronomical research we do at TSU concentrates on magnetism in cool stars. Specific projects are monitoring highly spotted variable stars, discovery of new variable stars, determining times of minimum for eclipsing binaries, interpreting eclipses of spotted (RS CVn) stars, monitoring semiregular variable stars, constructing chromospheric models for giant and supergiant stars, and defining the subtle luminosity changes in old Sun-like stars. Many of these are based primarily on photometry from APT's. However, most depend on using <u>combinations</u> of photometry and spectroscopy, with spectroscopy becoming increasingly important to us because spectra contain so much more information than individual photometric data. Our experience teaches that a combination of very efficient photometry from APT's with inefficient spectroscopy from manual observing runs at the National Observatories is quite limiting. We therefore yearn to do the spectroscopic parts of our research as efficiently as the photometry by building and managing an automatic <u>spectroscopic</u> telescope (AST). To preserve the simplicity required for reliability, such a telescope would necessarily be tailored to our research interests.

4. A Research Program for an Automatic Spectroscopic Telescope

We have identified a group of projects that form the basis of an observing program for an AST. This group includes a wide range of monitoring projects that would be either impracticable or prohibitively expensive when done with manual observing, several large-scale surveys of cool stars, and projects that improve the way data are taken and interpreted. These potential research projects are the following:

1. **Search for magnetic cycles in moderately to highly chromospherically active stars.** This program consists of monitoring stars with spectra from AST's at Hα and Ca II to determine how the line emission tracing chromospheric activity changes with time. Whether such chromospheric emission can change is unclear, but Dorren & Guinan (1990) have claimed an effect in HR 1099. Such spectroscopic monitoring is even more impracticable with manual observing than is the photometric monitoring.

2. **Routine Doppler images of spotted stars.** We still do not know what changes in the mean brightness of stars like RS CVn mean in terms

of magnetic cycles. Most of these highly active systems at any one time have two dark regions on their surfaces, which we call spots, and which persist for roughly two years each (e.g., Henry et al. 1995). They seem to be either persisting groups of moderately large spots or groupings of spots that just seem to persist. We intend to determine what causes the changes in effects of these spot groups by combining spectroscopy with photometry to define the changing spot distributions in a group of about five spotted stars. The advantage of an AST lies in the vastly better data base it can amass, which would allow both excellent phase coverage at individual epochs as well as several independent analyses for a given year.

3. **Variability of chromospheres in cool giant stars.** Little is known about how the chromospheric structures of cool giants, the ones with large mass loss in winds, change with time. Do they have magnetic cycles? Are there changes related to pulsation of the star? If so, there would be hints about the chromopspheric heating. We are obtaining exploratory observations in this area at NSO, but few stars can be studied in this way.

4. **Winds of cool giant stars.** Properties of winds in cool (G-M) giants are based primarily on a few observations of a moderate number of stars. We do not really know how winds in these stars change with time and what relation any such changes have to variations in chromospheric or photospheric structure. Many observations, over long timescales, are required to answer this question.

5. **Pulsations in SR variables and long-period variables.** These are M giants that pulsate multiperiodically on timescales of a few weeks to 200 days. UV observations show that their chromospheric emission is modulated on pulsational timescales in many cases, raising the possibility that chromospheric heating in these stars comes mostly from long-period waves (e.g., Eaton, Johnson, & Cadmus 1990). Intense monitoring of several in chromospheric lines is needed to explore this possibility.

6. **Variation of mass loss in R CrB stars and shell stars.** These are stars with high rates of mass loss thought to be passing from the stage of red supergiant to that of planetary nebula. An aspect of this mass loss is the formation of dust clouds that sometimes block our view of the star. It seems exceedingly queer that a star would puff off a cloud of gas or dust. Effects of the clouds appear dramatically in photometry; we are interested in the conditions in winds that lead to their formation, which may appear in spectra of wind lines.

7. **Eclipses of ζ Aur binaries.** Systems in which a K supergiant eclipses a B dwarf with its chromosphere and wind let us determine in detail properties of chromospheric gas in actual stars. Although UV spectra follow the chromosphere and wind to much greater height (e.g., Eaton & Bell 1994), optical observations may be used to sample the gas in the lower parts of the chromosphere where most of the intrinsic emission lines are formed. In fact, repeated observations of three systems, 31 Cyg, 32 Cyg, and ζ Aur, will be needed to define properties of the lower chromospheres of

cool supergiants. AST's promise the flexibility to make these observations practical.

8. **Orbits of long-period binaries and triple systems.** These stars may be analyzed to provide basic information about stellar masses and evolution. This is especially important for combination with measurements of these same stars from the coming generation of powerful interferometers.

9. **Mass-exchange events in Algol binaries.** These systems transfer mass from a less massive to a more massive component through Roche-lobe overflow. The mass exchange seems episodic and elusive, hence difficult to detect with manual observing. Statistical studies requiring many spectra over long periods of time will determine the rate of evolution in Algol systems.

10. **Mass ratios of Algol binaries.** Doppler images of these systems will give a measure of the sizes of the cool, mass-losing star, hence of the mass ratio of the system (e.g., Eaton, Hall, & Honeycutt 1991). Such information is needed to determine accurately the evolutionary stage of the system and to determine accurate masses and other properties of the stars for theoretical simulations.

11. **Survey of Hα in all bright cool stars.** Hα is a probe of the upper chromosphere in cool stars. Variations of it from star to star give an indication of the chromospheric variation expected in a given star. We have observed a significant fraction of the cool giants brighter than V=6 with the McMath-Pierce solar telescope (Eaton 1995). We intend to survey all the cool giants within reach of our telescope to determine how chromospheric absorption depends on metallicity as well as on basic properties of the star.

12. **Discovery of more stars with composite spectra.** These would be discovered by comparing Hα and violet spectra of a great many stars. They will be used to discover more long-period binaries and to assess differences in the chromospheric activity of single stars and binaries.

13. **Survey of Li strengths (abundances) in cool giant stars.** This is to tell better how Li abundance changes with evolutionary stage. Since Li depletion is still rather poorly understood theoretically, especially in binaries, an assay of as many stars as possible is needed to understand fully what abundances are really encountered under different conditions.

14. **Determination of how to obtain more precise routine spectra.** In order to accomplish our scientific goals, we must figure out just how to control and characterize the instrument to give reliable observations routinely. We have done this to dramatic effect with photometry. Doing so with spectra would vastly improve the quality of observational results and conclusions.

15. **More efficient analysis of spectra.** The data rate of an AST will force analysis programs to become more efficient in the way spectra are used. Thus an automatic spectroscopic telescope has the potential to revolutionize the way data are analyzed and interpreted and the results published.

Acknowledgments. I would like to acknowledge numerous helpful discussions of the ideas in this paper with Gregory W. Henry. This research has been supported in part by NSF grant HRD 9104484.

References

Dorren, J. D., & Guinan, E. F. 1990, ApJ, 348, 703
Eaton, J. A. 1995, AJ, in press (April).
Eaton, J. A., Hall, D. S., & Honeycutt, R. K. 1991, ApJ, 376, 289
Eaton, J. A., Johnson, H. R., & Cadmus, R. R. 1990, ApJ, 364, 259
Eaton, J. A., & Bell, C. 1994, AJ, 108, 2276
Eaton, J. A., Henry, G. W., Bell, C., & Okorogu, A. 1993, AJ, 106, 1181
Henry, G. W., Eaton, J. A., Hamer, J. W., & Hall, D. S. 1995, ApJS, in press

A proposal for a fiber-fed échelle spectrograph for a southern-hemisphere robotic telescope

J.B.Hearnshaw

Dept. of Physics and Astronomy, University of Canterbury, Christchurch, New Zealand

Abstract. A proposal is made for a high resolving power fiber-fed échelle spectrograph on a dedicated southern hemisphere robotic 2-m telescope, or for a network of several such telescopes. The astrophysics that might be accomplished is outlined and the possible design features of such a spectrograph are discussed. The purpose of the paper is mainly to stimulate interest in the merits of such a project.

1. Introduction

At the present time a fully automated 2-m class telescope, or network of such telescopes, in the southern hemisphere, feeding a high resolution CCD spectrograph via an optical fiber and dedicated to spectroscopy, would become a valuable workhorse for spectroscopic studies of southern variable stars to about $m_V = 10$ or perhaps 11. Since there are about 10^6 stars brighter than this limit at southern declinations, few of which have ever been observed at high resolution, a huge quantity of new data on interesting objects would become available. Studies of variable stars would be the central project for such a telescope. With a very stable instrumental profile and a resolving power of $R \sim 5 \times 10^4$, studies of line asymmetries, macroturbulence, stellar rotation and high precision radial velocities would become possible for many stars. These could include Mira variables even at minimum light, giants in nearer globular clusters, Magellanic Cloud supergiants, spectroscopic binaries, active chromosphere stars and programs undertaken could include the search for planets and brown dwarfs using Doppler techniques.

To achieve such goals a fiber-fed échelle spectrograph with no moving parts and equipped with a large format CCD could meet the required specifications. The purpose of this paper is simply to outline some of the design features of such an instrument which are considered to be desirable, in the hope that this will stimulate interest in constructing one or more such automated spectroscopic telescopes in the southern hemisphere.

This paper is based in part on the author's experiences since 1988 with a fiber-fed échelle spectrograph on the 1-m telescope at Mt John University Observatory in New Zealand (Kershaw and Hearnshaw 1989, Murdoch et al. 1993). Our present instrument is not an automated system, the spectrograph delivers a resolving power of $R \sim 31\,000$ with a 100 μm slit, and the small format CCD gives limited spectral coverage. The collimator beam of this spectrograph is 54 mm in diameter, not nearly large enough for the efficient use of stellar photons (Hearnshaw 1977).

2. Pros and cons of fiber-fed échelles

The advantages of fiber-fed échelles are well-known (see for example Heacox (1988)). They allow the spectrograph to be mounted off the telescope in a chamber with high thermal stability (temperature fluctuations of 0.01°C over a night are a reasonable goal), flexure problems are eliminated and light scrambling in the fiber avoids the influence of guiding errors on the instrumental profile.

On the other hand, the light throughput of fibers is more seeing-sensitive than a long and essentially one-dimensional rectangular slit, and fibers may have poor ultraviolet transmission. Focal ratio degradation may further worsen the spectrograph's performance, and in general it is difficult or impossible to match the fiber diameter to both the seeing at the input end and the required resolution at the fiber output. Although image slicing of the fiber output can help this problem, an image-sliced system requires more cross dispersion to separate orders and hence gives less wavelength coverage for a given detector format. Hence it is not advocated here.

For these reasons compromises must be made which might well degrade light throughput by about a factor of two relative to a telescope-mounted spectrograph (the actual figure depends on the seeing). Such a sacrifice is often warranted if high resolving power and stability are deemed to be desirable characteristics of the instrument.

3. Initial specifications

In this preliminary design study it is assumed that a dedicated 2-m automated alt-az telescope will feed a fiber at the f/6 Cassegrain focus (scale 14 arc s mm^{-1}). In addition $R = 50\,000$ is the resolving power design goal, as many interesting problems in stellar spectroscopy then become accessible. The instrumental profile should be as stable as possible, and for ease of automated operation, no moving parts or observer-made mechanical adjustments should be provided in the spectrograph, except possibly an option for a wider slit and lower resolving power to provide higher throughput on fainter stars.

The standard equation for a slit spectrograph's resolving power is:

$$R = 2L \sin \theta_B / (\theta_S D)$$

or, in those cases where the échelle might be overfilled by the collimator beam:

$$R = 2f_{\text{coll}} \tan \theta_B / W = 2B \tan \theta_B / (\theta_S D)$$

Here $L =$ size of échelle grating illuminated (perpendicular to grooves)
 $\theta_B =$ échelle blaze angle (assumed to be near-Littrow configuration)
 $B =$ beam size from collimator (may overfill échelle)
 $\theta_S =$ angular size of slit (radians)
 $D =$ telescope aperture
 $W =$ slit size

The corresponding equations for a fiber-fed spectrograph are:

$$R = 2L \sin \theta_B / (\rho \theta_F \epsilon_S D)$$

(if the collimator beam does not overfill the échelle) or

$$R = 2f_{\text{coll}} \tan \theta_B / W = 2B \tan \theta_B / (\rho \theta_F \epsilon_S D)$$

(here the collimator beam may overfill the échelle),

where θ_F = angular size of the fiber input (radians)
ρ = focal ratio degradation factor (≥ 1)
$\epsilon_S = W/F$, the ratio of slit size to fiber diameter, such that $\epsilon_S \leq 1$.

Clearly high resolving power still requires a large échelle grating, L, and in particular, a large collimator beam size, B. The merit of a spectrograph M can be defined as the product of resolving power R and light throughput T ($M = RT$). Since throughput $T \propto D^2$ and resolving power $R \propto B/D$, it follows that $M \propto BD$. Thus large beam size B is just as beneficial as large telescope aperture D to overall spectrograph performance. Often the cost of increasing beam size will be less than that of a larger telescope. A simple analysis of costs shows that if merit is to be maximized for a fixed number of dollars available for a telescope plus spectrograph system, then as much should be expended on the spectrograph, so as to achieve the largest possible beam size, as on the telescope.

In practice this means that the spectrograph with the largest possible échelle should be built, which is presently a Milton Roy 204 × 408 mm échelle.

4. Design requirements

It would be desirable to satisfy all the following design requirements in the spectrograph design:

1. Match fiber input (θ_F) to seeing (θ_*)

2. Match fiber output diameter (F) to slit size (W)

3. Match the image of the slit on the detector (W') to the detector pixel size (p), so that $W' = 2p$

4. Match the collimator beam size B to the projected dimension of the échelle ($L \cos(\alpha + \theta)$)

5. Match the free spectral range length of the orders (Δl_{FSR}) to the size of the CCD detector (P)

6. Achieve the required resolving power $R \sim 50\,000$

The available parameters to vary in the design are:

1. The fiber core diameter F (this also fixes the angular size of the fiber entrance on the sky (θ_F))

2. The collimator focal length (f_{coll})

3. The camera focal length (f_{cam})

4. The slit size (W)

5. The overall CCD dimension (along the orders) (P)

Since there are fewer variables than design criteria to satisfy, a general solution is not possible. This is why compromises are essential in spectrograph design. For the instrument being considered here, it is essential that the resolving power not be compromised, as this will affect the astrophysics that can be done. Nor should the wavelength coverage be sacrificed by allowing the échelle orders to be larger than the CCD detector, because an automated instrument will not have the versatility of a tiltable échelle grating or cross disperser to change the recorded wavelength region. Furthermore, the Nyquist sampling theorem should not be violated, as doing so rapidly degrades the overall merit or the wavelength coverage. On the other hand, compromising requirements 1, 2 or 4 above can slightly reduce throughput, yet give larger percentage gains in resolving power than loss in throughput, thereby maximizing the merit.

Here it is therefore proposed that:

a) A 150 μm fiber should be used, equivalent to a 2.1 arc s circular aperture centered on a star. This will accept 42 per cent of the light in an image in 2.0 arc s seeing or 79 per cent in 1.0 arc s seeing (Diego 1985). A larger fiber (say 200 μm) accepts more light but is less well matched to the required resolving power.

b) The fiber exit should be reimaged onto a 122 μm-wide slit (no magnification), so as to improve the resolving power.

c) The collimator beam is arranged to overfill the échelle. The beam exiting from the fiber is assumed to be at focal ratio f/5 due to focal ratio degradation ($\rho = 1.2$). The beam size of $B = 300$ mm from a 1500 mm focal length collimator is larger than the size of the échelle seen in projection from the collimator (208 × 157 mm). The large collimator focal length improves resolving power (Tull 1972) ($R \propto f_{coll} \propto B$), but the light loss is not necessarily severe, especially as there will be a low intensity in the peripheral rays of the f/5 beam from the fiber.

Other proposed design features are simply noted briefly as follows:

d) The Littrow angle θ (the difference between the angle of incidence on the échelle (α) and the blaze angle, (θ_B) should be small. $\theta = 6°$ might be typical in cassegrain échelle spectrographs, but $\theta = 4°$ should be the target. This improves échelle efficiency, but results in a larger spectrograph if vignetting is to be avoided. However overall instrument size is not a primary consideration.

e) A 31.6 grooves/mm R2 ($\theta_B = \arctan 2$) échelle should be used. The commercially available values are 31.6, 52.65 and 79.0 grooves/mm. But the coarser rulings give shorter order lengths, thus enabling full spectral coverage to the far red on a large format CCD.

f) A 2048 × 2048 pixel CCD with square format of 50 mm (24 μm pixels) would enable orders from $n = 148$ ($\lambda = 380$ nm) to $n = 65$ ($\lambda = 868$ nm) to be recorded with at least one free spectral range being on the available 50 mm detector size.

g) Cross dispersion using two large 48° apex angle prisms in Schott UBK7 glass would allow 86 orders ($n = 63$ to 148, $\lambda = 380$ to 900 nm) to be simultaneously recorded with complete wavelength coverage, except in the longest two orders ($n = 63, 64$). The mean interorder spacing would be about 580 μm, whereas the order width would be 150 μm.

h) A Schmidt-Cassegrain f/2.8 camera of 815 mm focal length would give a dispersion of 2.0 Å/mm at Hα and 1.3 Å/mm at Hγ, and satisfy the Nyquist sampling theorem of two pixels per resolution element.

5. Further proposed spectrograph features

An automatic telescope and spectrograph require an exposure meter, which receives either undispersed zero-order light from the échelle, or perhaps better still, the light overfilling the échelle could be collected and sent to an integrating photomultiplier photometer, for automatic exposure termination once sufficient signal has been received, or once cloud has reduced the rate of arrival of signal to zero for a certain interval of time.

Quality checks on the system will be needed each night to verify good spectrograph focus (using the Th/Ar comparison line width on the CCD) and good light throughput, by monitoring the light of the white flat-field lamp spectrum on the CCD. Both these lamps are housed in the fiber-feed module, and, when selected, they are automatically arranged to send light at focal-ratio f/6 into the fiber in the same way as starlight.

Since one desirable astrophysical goal is a very stable instrumental profile (or point spread function), a sealed and evacuated spectrograph would give immunity to refractive index changes of the air due to air pressure and temperature variations. Note that a 1 mbar pressure increase mimics a Doppler shift of -80 m/s; a 1°C temperature change mimics a shift of +270 m/s, in each case the wavelength change arising from a change in the air's refractive index. A sealed spectrograph gives immunity to air pressure variations, whereas an evacuated spectrograph in addition avoids image shifts arising from convective currents and ray refraction in stratified air layers. An evacuated spectrograph would also facilitate the maintenance of clean optical surfaces. Thus a large evacuated tank at a pressure of about 0.1 mbar would ensure adequate protection from the vagaries of the atmosphere. The temperature of the tank and its contents would be controlled to about 0.01°C on time scales of several hours.

The CCD liquid-nitrogen dewar would be designed to vent to outside the room containing the spectrograph tank, and filling of the dewar would be fully automated on a daily basis.

6. The case for southern hemisphere automatic spectroscopic telescopes

An excellent case can be made for a network of three or four 2-m class automated spectroscopic telescopes in the southern hemisphere, possibly at sites in Chile, South Africa, Western Australia and New Zealand, so as to give continuous coverage of southern variable stars. The longitudes are about 70°W, 20°E, 116°E and 171°E respectively. There are almost certainly benefits from operating at established observatory sites, so that regular maintenance of the telescope and instrumentation can be undertaken. The New Zealand site (Mt John University Observatory, latitude 44°S) offers the additional advantage of year-round visibility of the Magellanic Clouds, which is difficult from sites at lower latitudes.

The intention of the present paper is no more than to stimulate interest in such a project. Those readers who are incited to explore the feasibility of this challenge further are invited to communicate with the author (Internet: phys012@csc.canterbury.ac.nz).

References

Diego, F. 1985, PASP 97, 1209
Heacox, W. D. 1988, Astron. Soc. Pacific Conf. Ser. 3, 204
Hearnshaw J. B. 1977, Proc. Astron. Soc. Australia 3, 102
Kershaw, G. M. & Hearnshaw, J. B. 1989, Southern Stars 33, 89
Murdoch, K. A., Hearnshaw, J. B. & Clark, M. 1993, ApJ 413, 349
Tull, R. G. 1972, ESO/CERN Conf. on Auxiliary Instrumentation for Large Telescopes, p.203

NETWORK OF ORIENTAL ROBOTIC TELESCOPES

FRANCOIS RENE QUERCI
Observatoire Midi-Pyrénées, 14 Av. E. Belin, 31500 Toulouse, France

MONIQUE QUERCI
Observatoire Midi-Pyrénées, 14 Av. E. Belin, 31500 Toulouse, France

SAMIR KADIRI, C.N.C.P.R.S.T., 52 Charii Omar Ibn Khattab, Agdal, B.P. 1346 R.P., Rabat, Royaume du Maroc

ZOUHAIR BENKHALDOUN
Dépt. de Physique, Faculté des Sciences Semlalia, Univ. Cadi Ayyad, B.P. S 15, Marrakech, Royaume du Maroc

and the Engineers of Observatoire de Haute-Provence, C.N.R.S., O4870 Saint-Michel l'Observatoire, France

ABSTRACT   A project to establish a network of automated photometric astronomical telescopes for observing variable stars, the ORT network, on high mountains having high-quality astronomical conditions in interested arabic and islamic countries is under progress. The data are to be automa-tically shared via telecommunication satellites with all the members of the network.

INTRODUCTION

We should like to report on the Network of Oriental Robotic Telescopes, otherwise the ORT network. Our objective is non-stop observations of variable stars by collaborating with other networks of automated photometric telescopes which are complementary to the ORT network in longitude and latitude intervals, such as the GNAT in USA, the chilean cordillera stations, the South Africa stations, etc. In a first step we will deal with photometric telescopes, then we will extend them to spectroscopy.

SITE SELECTION

Using 12-year archives of meteorological satellites, it appeared that sites located around the north tropical latitude from 15° to 35° and from 10° West to 110° East in longitude have high-quality astronomical conditions with a significant annual number of clear nights. Such sites involve arabic and islamic countries from Morocco to the western deserts of China.

In addition to their astronomical history, these countries are suitable because they have high mountains in semi-desertic areas, i.e. a clean sky with low telluric absorption.

The final site selection will be based upon local astronomical tests, such as seeing, scintillation and telluric absorption measurements.

The meteorological prospecting together with the local access facilities ought to give a list of network sites not subjected to the same airstreams. The minimum number of the network stations able to follow variable stars each night without interruption, will be defined around 10 to 12.

PHOTOMETRIC TECHNIQUES

As for the photometric technique, we advocate differential photometry which, of course, implies a comparison of the flux of the variable star, Va, with the flux of two (or more) comparison stars, A and B, known to be non-variables and which are located in the neighbourhood of Va. To freeze the sky transparency changes during the observation, we develop techniques allowing to observe the three stars Va, A, B and the sky simultaneously (in using many telescopes in the same site -- Querci et al., 1992, and references herein -- or one large field telescope with a battery of CCD cameras). This leads to an improved accuracy and a saving of the observing time by minimizing the telescope maneuvers and by removing the successive observations of the comparison stars.

The high accuracy level of the simultaneous observations of the three stars and the sky is reached even when the sky conditions are slowly variable.

The final choice between the multiple telescope technique with photomultipliers as receivers and the single large field telescope with a battery of large CCD cameras, depends on many factors, such as the type of research projects foreseen, the technical advantages of each technique, and the relative cost of each technique.

SHARING OF THE DATA

The data collected by each station will be automatically and simultaneously transmitted by telecommunication satellites to all the scientific centers of the network, making common data reductions possible.

STATUS OF THE PROJECT

The design of an entire typical stations : weather unit, telescopes, photometers and antenna for satellite communications etc., has been done at the Observatoire de Haute-Provence in France. There, the engineers have analysed the advantages and the disadvantages (reliability, etc.) of each of the various ways which permits simultaneous star observations. After having tested the multiple telescope solution with small telescopes, a 1-m diameter Ritchey-Chrétien telescope with a large field (1°20') and a battery of CCD cameras will be soon under test.

TEACHING AND TRAINING SCHEMES

Included in the project are teaching and training schemes on the astronomical technology used in the network which are supplied by Observatoire de Haute-Provence (OHP), and also basic courses on astrophysics and student practice on an equiped small telescope (60-cm diameter) to be given in Universities of arabic and islamic countries who ask them.

THE NEXT WORKSHOPS

The scientific and/or the technical participation will be discussed during local or regional workshops. The first is planned to be held in September 1994 in Jordan.
In addition to the teaching organization, will be discussed questions such as :
- local site testing campaigns, - cooperative scientific programmes to be developed with the network and consequently the diameter of the telescopes, - choice of the first and second generations of focus equipments (photometers, spectrographs, etc.), - practical organization of the network board, - financial support by the interested countries and international organizations, - etc.

CONCLUSION

The ORT network, a network of robotic telescopes, will complement the automated stations located in other longitude and latitude intervals, allowing large international collaborative variable star programmes to be run in a non-stop way for several nights or weeks.

This project is partly supported by "Institut National des Sciences de l'Univers" (INSU/CNRS) and by "Département des Alpes de Haute-Provence" where OHP is located.

REFERENCES

Querci, F.R., Querci, M., Kadiri, S., de Rancourt, L. 1992, in *Proceedings of the IAU Colloquium 106, Poster Papers on Stellar Photometry*, ed. I.Elliott and C.J. Butler, p. 122.

Astronomy From The Moon

Jack O. Burns
Department of Astronomy, New Mexico State University, Las Cruces 88003

Abstract. In this paper, I ask the question: What unique astronomical observations should we plan to do from the Moon? I discuss four possible telescopes that make good use of the lunar environmental conditions, that will contribute to a fundamental new understanding of astrophysical systems, and that will begin with soft-landed, robotic precursors. These include radio telescopes on the lunar far-side, a thermal infrared telescope within a permanently shadowed crater at a lunar pole, a 10-km baseline optical/IR imaging interferometer, and a high throughput X-ray telescope. I will also discuss new lunar environmental data that must be gathered before any of these large systems are placed on the Moon.

1. Introduction

These conference proceedings contain many interesting and innovative ideas for automated photometry, CCD imaging, spectroscopy, and even optical/IR interferometry. Clearly, the technology of robotic telescopes is advancing quickly and this is beginning to produce first-rate science. However, it is equally important to locate our telescopes at the best possible sites if we are to obtain the very best astronomical data. This has been superbly illustrated with the CARA observatories in the Antarctic where the high altitude and low water-vapor content of the atmosphere is superior to any other site on the Earth, especially for infrared and submillimeter observations. The more remote the site, the greater the need for robotic telescope operations.

Possibly the best site for advanced, robotic and manned, telescopes is the surface of the Moon. The advantages of the Moon for astronomical observations are now well known (see e.g., Burns & Mendell 1988; Mumma & Smith 1990; Burns 1992) – the radio-quiet far-side, the negligible atmosphere, the two week long nights, the geologic and orbital stability, and the abundant surface area of the Moon.

The best astronomical sites also cost more to operate due to the limited access and generally harsh conditions. The Antarctic observatories are a factor of 5-10 more expensive than similar observatories on the mainland U.S. The Moon, which is about three days away by spacecraft, will probably be 10-100 times more costly for astronomical telescopes of a given aperture than Earth-based systems. So, although the potential science from these remote sites is excellent, the cost is also high. Thus, one must carefully choose which telescopes to locate at these sites so as to produce truly major advances in astrophysics and justify the great costs. The search for the best possible science from the Moon will be among the themes of this paper.

In the case of the Moon, it is clear to me that we will someday return to its surface with human bases and astronomical telescopes. The Moon is too important a stepping stone toward exploration of the solar system, given its proximity, to ignore for very long. Once a human base is established and regular transportation to the Moon is present, the cost of doing science from the Moon will drop. At that time, it will be feasible to consider some of the imaginative "grand" observatories that have been proposed for the Moon (see e.g., Burns et al. 1988; Mumma & Smith 1990). For the next decade or two, however, national and international budgetary pressures require that only modest, robotic telescopes on the lunar surface are feasible. We should continue to plan for the larger-scale lunar telescopes via careful design of unmanned, soft-landed telescopes which will serve as crucial precursors. There is also much about the lunar environment (e.g., dust mobility, micrometeoroids, cosmic rays) that is still poorly known, yet such data are crucial in determining the engineering and observational constraints of future lunar-based telescopes.

In this paper, I will concentrate on four possible telescope projects for the Moon which, in my judgement, make the best use of the unique natural lunar resources. I have chosen these four telescopes based upon a survey of the now-abundant literature (primarily conference proceedings) on lunar-based telescopes and some consideration of balance across the wavelength spectrum. I will describe the attractive lunar attributes for these telescopes, the basic properties of the "ultimate" telescopes, science highlights, and robotic precursors which are needed to gather engineering and science data for the grand lunar-based observatories.

2. Radio Astronomy from the Lunar Far-Side

The far-side of the Moon is the only location in the near-Earth environment that has been demonstrated to be radio-quiet. The RAE-2 satellite, which orbited the Moon in the early 1970's, showed that the Earth is a major source of radio interference especially below 10 MHz. In Figure 1, from Alexander et al. (1975), we see that the recorded antenna temperature of the sky is that expected from the Galactic background when RAE-2 was above the lunar far-side, but far higher when the Earth was in view. From 1-10 MHz, ionospheric breakthrough of human-made interference and terrestrial atmospherics (i.e., lightning) are the major sources of interference. Below 1 MHz, the Earth's magnetotail was discovered by RAE-2 to be an unexpectedly strong source of radiation probably caused by plasma maser instabilities. Of course, observations at < 10 MHz from the Earth's surface are difficult or impossible because of the large optical depth of the ionosphere. So, the lunar far-side is ultimately the best location to open this last window to the electromagnetic spectrum at very low radio frequencies as noted in the Bahcall Committee report (Bahcall et al. 1991).

The community of low frequency astronomers has proposed a strategic plan for space-based low frequency radio astronomy as summarized in Table 1 (see also Kassim & Weiler 1990). This evolutionary approach begins with a simple frequency-swept spectrometer in low Earth orbit, OHFRIM, which will defini-

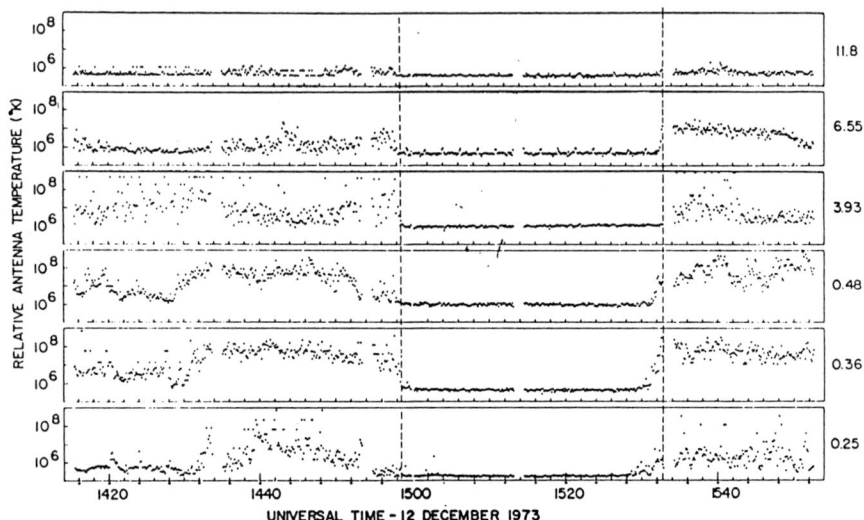

Figure 1. Very low frequency emission as observed from lunar orbit by RAE-2 (from Alexander et al. 1975). Relative antenna temperature is plotted for various receiver frequencies (right hand ordinate in MHz). Note the sharp drop in noise when the Moon occulted the Earth between 1500 and 1530 UT.

Table 1. Strategic Plan for Space-Based Low Frequency Radio Astronomy

Phase	Instrument	Capability
1	OHFRIM[a]	Spectral analysis on a single spacecraft
2	TELFI[b]	Low dynamic range mapping of bright sources
3	Orbiting Multielement Array	Aperture synthesis using closure amplitude & phase
4	Far-Side Array	High dynamic range imaging, excellent (u,v) coverage

[a]Orbiting High Frequency Radio Interference Monitor
[b]Two-Element Low Frequency Interferometer

tively measure for the first time the low frequency spectrum (0.5-16.5 MHz) of radiation emanating from the Earth (Kassim, Weiler, & Clegg 1994). This will determine if there are any windows or times where low frequency astronomy can be performed in Earth orbit. The next mission, TELFI, is a two-element interferometer that would orbit the Moon or the Earth (depending on the results from OHFRIM) and construct the first arcsecond resolution images with modest dynamic range of the few hundred brightest sources in the sky (Burns, Basart, & McCune 1994). Such mapping will be possible using a new design, inflatable, phased-array antenna on each satellite (Basart, Mandayam, & Burns 1994). Such a three dimensional antenna has a mass of only a few kilograms, requires no moveable parts, and has a directivity that is 200 times better than a dipole antenna (thus reducing the source confusion in the beam and greatly improving the sensitivity). I believe that this technology may have applications to high-gain space-based antennas at microwave frequencies. TELFI will be capable of making 1-2 arcsec resolution maps for high Galactic latitude sources, comparable to the VLA at 20-cm (Burns 1991). These precursor telescopes would then be followed by the "ultimate" array composed of hundreds of elements distributed over large areas on the lunar far-side, producing high dynamic-range, relatively high resolution images of the low frequency sky (see Burns, Johnson, & Taylor 1989).

One of the most exciting science drivers for these very low frequency radio telescopes involves the origin and evolution of cosmic rays (Duric & Burns 1994). Only a relatively small portion of the energy spectrum of cosmic rays is observed on the Earth. Modulation of the lower energy (and more abundant) cosmic rays by both the Earth's magnetosphere and the solar wind magnetic field means that these particles never reach the Earth's surface. Yet, the shape of the energy spectrum below 1 GeV is a critical constraint on cosmic ray acceleration theories. More generally, cosmic rays significantly influence the ISM of the Milky Way and possibly star formation. However, 200 MeV electrons within 100 μG magnetic fields, typical of supernova remnants, emit 13 MHz photons via synchrotron radiation. So, very low frequency observations with TELFI or a far-side array could provide key data on particle acceleration mechanisms in the ISM of our Galaxy.

In addition to very low frequency observations, the lunar far-side may someday become important for microwave telescopes. Human-made interference within "radio-quiet" bands is an increasing problem. Marked yearly increases in interference at the crucial 21-cm band at the VLA is now well documented (Perley 1994). A variety of causes include new satellite transmissions, GPS satellite sidelobes, radar, and military operations. At this rate, in 25-50 yrs we may not be able to operate at the 21-cm band from the Earth's surface. Drake (1988) has suggested placing Arecibo-style antennas within lunar far-side craters for very high sensitivity microwave observations. He suggests that dishes with 30-km diameters (using conventional building materials) may be possible on the Moon because of the low gravity. Such antennas would provide extremely sensitive observations of HI in distant spiral galaxies. These telescopes will, of course, only be successful if we actively work to prevent radio noise pollution on the lunar far-side.

3. Thermal Infrared Observations

Lester (1988, 1992) has made a very interesting proposal for a thermal infrared telescope that would be sited within a crater at one of the lunar poles. This is now particularly timely because of recent discoveries by the Clementine spacecraft which orbited the Moon.

Lester (1992) proposed that lunar polar craters are ideal sites for infrared telescopes which would operate over the 10-50 μm band. The DoD Clementine mission has recently confirmed that there are craters near the poles which are permanently shadowed with equilibrium temperatures possibly as low as 25 K. Furthermore, there may be evidence of ancient water ice within these craters from the Clementine backscatter radar data. If this is confirmed, the lunar polar craters will be obviously high profile targets for further robotic and human explorations. The lunar poles would be the only source of water on the Moon and possibly a key to future lunar colonies.

Thermal infrared telescopes require cooling of not only the detector but also the optics and the telescope superstructure. Passive cooling would come for free inside one of the shadowed lunar polar craters, thus greatly simplifying the design and reducing the cost of the telescope. Such a telescope might be carried as an add-on payload to a future robotic lander mission which is intended to survey these craters and its water ice content.

Lester proposes that the Edison telescope (Thronson et al. 1990) could serve as in important technology precursor to a thermal infrared lunar telescope. Edison is a 2-m telescope that would fly in high Earth orbit and would be passively cooled to 30-40 K. Lester suggests that the passive cooling technology needed for the lunar polar telescope could be developed by Edison and transferred to his telescope.

Lester is somewhat vague on the details of his proposed lunar polar crater telescope. However, a first telescope with an aperture of 1-2 m, depending on the availability of light-weight optics, would do a very nice job of performing a deep sky survey in the thermal infrared.

The science from such a telescope is quite obvious and important. One such focus would be on star formation, young stellar objects, stellar jets, and protoplanetary disks. The far-infrared offers important insights into the energetics of star-forming regions, including probes of molecular and atomic coolants. This is one of the most exciting and quickly advancing topics in modern astrophysics.

4. Optical/IR Interferometry

The one proposed lunar-based telescope that has garnered the most attention is an optical/IR imaging interferometer (see e.g., Burke 1990). It may be the greatest leap in technology, resolution, and science of any proposed telescope for the Moon. The Moon is a particularly suitable site for this telescope because of the large, relatively flat area that is needed for a multi-kilometer baseline array, the very low optical depth atmosphere, passive cooling available for the optics during the lunar night, and the relatively stable baselines. Such a telescope has been discussed by Burke (1990), Shao (1990), and Burns, Johnson, & Duric (1992).

Two designs have been proposed for this interferometer. The first is a Y-shaped distribution of elements, similar to the NRAO Very Large Array in New Mexico (Burke 1990). The 27 four-meter aperture elements would be distributed with a maximum baseline of 10-km producing an impressive resolution of 10μarcsec at 500 nm. This design has the advantage of combining three linear arrays making the beam combination relatively simple. Alternatively, a 1989 workshop held in Albuquerque recommended that the interferometer elements should be distributed along a nonredundant Cornwell circle with an outer diameter of 10-km (Burns et al. 1992). This has superior instantaneous (u, v) coverage which is important on the Moon given the slow rotation rate. However, combining the beams and dealing with the polarizations of the beams will be more complex.

The most exciting potential science with a lunar optical/IR telescope would be the detection and imaging of an Earth-like planet in another stellar system. Interferometric techniques can be employed to reduce the effects of the primary star and allow the detection of a Jupiter-like or even an Earth-like planet. The interferometer would use a beam combiner that nulls out the light from the star while passing light at 10μm from the planet (e.g., Shao 1990). Simulations suggest that such an experiment is well within the instrument parameter bounds for stellar systems at the distance of τCeti (3.6 pc).

Clearly, the demands on the technology and the site are substantial for a lunar optical/IR interferometer. Precursor telescopes are essential to better understand the environmental conditions on the Moon, to investigate the operation of an interferometer, and to refine the science goals. Thus, a simple 1-2 m class soft-landed, robotic telescope is an important first step toward an interferometer. To maximize the science, such a modest aperture telescope could operate in the far-UV (around 200 nm) where the sky is extremely dark and where few space-based observations have been performed to date. McGraw (1992) has suggested a 1-m lunar transit telescope (LUTE) which has few moving parts and would do a repeated transit survey of a strip(s) of the sky. Another possibility is a fully-steerable 1-m class telescope that would perform the first all-sky CCD imaging survey in the far-UV (Genet et al 1992). Either concept provides the necessary engineering data for a more informed design study of a future lunar interferometer, and itself will produce excellent science.

5. High Energy Astronomy on the Moon

Peterson (1990) has suggested that the Moon is a good location for high energy astrophysics telescopes that require long baseline systems (i.e., a large distance between the detector and the aperture) or high throughput. One such lunar telescope is LAMAR (Large Modular Array of Reflectors) proposed by Gorenstein (1990). This telescope has 10^4 cm^2 of collecting area, high throughput, high sensitivity (10^{-14} photons cm^{-2} sec^{-1}), excellent spectral resolution ($E/\Delta E \approx 300$), and modest spatial resolution (30 arcsec). Because of the size of this telescope, Gorenstein suggests that LAMAR is best built on the surface of the Moon using human technicians from a lunar base.

Gorenstein advocates using LAMAR as a cosmology telescope. The very high sensitivity of LAMAR will allow astronomers to detect X-ray emission from galaxy clusters and AGNs out to $z \approx 1$. The resulting statistical sample will cover a much larger volume than any current or planned optical redshift telescope. Gorenstein believes that LAMAR will allow us to finally probe a fair sample of the Universe with large enough scale sizes of structure to constrain dark matter models and possibly the value of Ω (ratio of average density of the Universe to the critical density).

As a precursor to LAMAR, a Lunar Transient Observatory (LTO) has been proposed (Peterson 1990). LTO would consist of high-energy γ-ray and X-ray detectors along with UV, IR, and optical CCDs. LTO would make simultaneous measurements over a wide range of wavelengths and over a large region of the sky. The optical and UV CCDs could be used to measure source positions to within an arcsecond. LTO is a relatively simple observatory that could be soft-landed on the Moon and remotely-operated from the Earth.

6. Summary & Conclusions

In Table 2, I summarize the properties of the four candidate astronomical telescopes that were discussed in this paper. It is a diverse set with a wide range

Table 2. Summary of Lunar-Based Telescopes

Lunar Driver	Candidate Telescopes	Science Highlight	Precursor
Radio Astronomy from the Lunar Far-Side			
Radio-quiet	Synthesis array	Acceleration of cosmic rays	TELFI
Negligible ionosphere	Arecibo-style crater telescope		
Thermal Infrared Observations			
Passive cooling to 25 K	IR sky survey	Star formation	Edison
No atmosphere			
Optical/IR Interferometry			
Ample real estate	VLA-like Y	Image Earth-like planet	LUTE
No atmosphere	Cornwell Circle		Far-UV sky survey
Stable surface & orbit			
Passive cooling at night			
High Energy Astronomy			
Long baselines possible	LAMAR	$z = 1$ structure	Transient Observatory
Stable base			
Long observing times			

of technologies, scientific goals, and expected costs. Each makes good use of the unique environmental properties of the Moon and each is strongly motivated by important astrophysical goals. In reviewing these telescopes, the following suggestions come to mind on how we might proceed:

1. We need to know much more about the properties of the lunar surface before large lunar-based telescopes are designed and built. For example,

a horizon glow toward the sunrise direction on the Moon was recently confirmed by the Clementine CCD cameras (H. Zook, private communication). The most likely interpretation of this glow is scattering of sunlight by dust that is > 1 m off the lunar surface. Criswell (1972) has suggested that dust is photoelectrically charged and accelerated off the lunar surface at the day/night terminator by ≈1000 V/cm electric fields established by charge separation over elevated lunar features. Doe et al. (1994) have found that this scenario is viable via a laboratory vacuum chamber experiment using lunar dust simulant. If dust is very mobile, this could create problems for optical surfaces and mechanical components. Similarly, the flux of micrometeorites and cosmic rays is poorly determined on the Moon. Each can have a detrimental effect on lunar-based telescopes. So, our group in New Mexico has proposed environmental monitoring experiments that could be carried on a first robotic lunar mission which would gather these crucial engineering data.

2. The use of robotic precursor telescopes will be essential to explore the technologies and the astronomical science that should be done from the Moon. Such missions should be planned and executed over the next decade. With careful cost-constraints and using light-weight optics, such robotic lunar telescopes can be flown for the same price as Explorer or Discovery class missions.

3. International collaborations are essential for lunar telescopes. Given the world economic conditions, it is very difficult for one nation to mount a systematic expedition to the Moon or even a series of robotic telescopes on the Moon. However, combining forces with the Russians, for example, and using their launchers and landers may provide a cost-effective means to fly these telescopes and to foster international cooperation.

4. Finally, it is my opinion that the so-called "grand" or ultimate lunar observatories cannot be built without a permanently occupied lunar base in place. Most of the large telescopes are too complex and too costly to assemble without an extensive human presence. Furthermore, a human base will provide the infrastructure, the transportation system, and will greatly reduce the costs of lunar telescopes (especially if building materials can be mined from the Moon rather than brought from Earth).

Acknowledgments. There are many people who have contributed ideas for the lunar-based telescopes discussed in this paper. I would like to thank my collaborators Steve Doe, John Basart, Stewart Johnson, P.W. Keaton, Jim Blacic, Don Petit, Neb Duric, Max Nein, Namir Kassim, Kurt Weiler, Bernie McCune, and Russ Genet for their hard work and input. I would also like to acknowledge fruitful discussions with Harlan Smith, Bernie Burke, Mike Shao, Dan Lester, Dan Weedman, Mike Ledlow, Mike Kaplan, and Paul Gorenstein. This work was partially supported by grants from the NASA Marshall Space Flight Center (NAG8-986) and from Autoscope.

References

Alexander, J., Kaiser, M., Novaco, J., Grena, F., and Weber, R. 1975, A&A, 40, 365

Bahcall, J. et al. 1991, The Decade of Discovery in Astronomy & Astrophysics, Washington: National Academy Press

Basart, J., Mandayam, S. & Burns, J.O. 1994 in Engineering, Construction, and Operations in Space IV, eds. R. Galloway & S. Lokaj, New York: ASCE, Vol. 2, 1390

Burke, B. 1990, Science, 250, 1365

Burns, J.O. and Mendell, W., eds., 1988, Future Astronomical Observatories on the Moon, NASA Conference Publication 2489

Burns, J.O., Johnson, S., & Taylor, J, eds., 1989, A Lunar Far-Side Very Low Frequency Array, NASA Conference Publication 3039

Burns, J.O. 1991 in Radio Interferometry: Theory, Techniques, and Applications, eds. T. Cornwell & R. Perley, San Francisco: ASP, Vol. 19, 420

Burns, J.O. 1992 in Robotic Telescopes in the 1990s, ed. A.V. Filippenko, San Francisco: ASP, Vol. 34, 273

Burns, J.O., Johnson, S., & Duric, N., eds., 1992, A Lunar Optical-Ultraviolet-Infrared Synthesis Array (LOUISA), NASA Conference Publication 3066

Burns, J.O., Basart, J., & McCune, B. 1994 in Engineering, Construction, and Operations in Space IV, eds. R. Galloway & S. Lokaj, New York: ASCE, Vol. 2, 1400

Criswell, D. 1972, Proc. Third Lunar Science Conf., Geochim. Cosmochim Acta, Suppl. 3, Vol. 3, Cambridge: MIT Press, 2671

Doe, S., Burns, J.O., Pettit, D., Blacic, J., & Keaton, P.W. 1994 in Engineering, Construction, and Operations in Space IV, eds. R. Galloway & S. Lokaj, New York: ASCE, Vol. 2, 907

Drake, F.D. 1988 in Future Astronomical Observatories on the Moon, eds. J.O. Burns & W. Mendell, NASA Conference Publication 2489, 91

Duric, N. and Burns, J.O. 1994 in Engineering, Construction, and Operations in Space IV, eds. R. Galloway & S. Lokaj, New York: ASCE, Vol. 2, 1382

Genet, R.M., Genet, D.R., Talent, D.L., Drummond, M., Hine, B.P., Boyd, L.J., and Trueblood, M. 1992 in Robotic Telescopes in the 1990s, ed. A.V. Filippenko, San Francisco: ASP, Vol. 34, 289

Gorenstein, P. 1990 in Astrophysics from the Moon, eds. M. Mumma & H. Smith, New York: AIP, 382

Kassim, N. and Weiler, K., eds., 1990, Low Frequency Astrophysics from Space, New York: Springer-Verlag

Kassim, N., Weiler, K., & Clegg, A. 1994 in Engineering, Construction, and Operations in Space IV, eds., R. Galloway & S. Lokaj, New York: ASCE, Vol. 2, 1410

Lester, D. 1988 in Future Astronomical Observatories on the Moon, eds. J.O. Burns & W. Mendell, NASA Conference Publication 2489, 85

Lester, D. 1992 in Robotic Telescopes in the 1990s, ed. A.V. Filippenko, San Francisco: ASP, Vol. 34, 325

McGraw, J.T. 1992 in Robotic Telescopes in the 1990s, ed. A.V. Filippenko, San Francisco: ASP, Vol. 34, 305

Mumma, M. and Smith, H., eds., 1991, Astrophysics from the Moon, New York: AIP

Perley, R.A. 1994, Very Large Array Observational Status Summary, Charlottesville: NRAO

Peterson, L. 1990 in Astrophysics from the Moon, eds. M. Mumma & H. Smith, New York: AIP, 345

Thronson, H.A., Hawarden, T.G., Mountain, C.M., Davies, J.K., Lee, T.J., and Longair, M. 1990 in Observatories in Earth Orbit and Beyond, ed. Y. Kondo, Dordrecht: Kluwer, 501

Shao, M. 1990 in Astrophysics from the Moon, ed. M. Mumma & H. Smith, New York: AIP, 486

A PRACTICAL AND AFFORDABLE TELESCOPE FOR THE MOON

PETER C. CHEN
Astronomy Program, Computer Sciences Corp., CSC/
NASA Goddard Space Flight Ctr., Greenbelt, MD 20771

R. J. Oliversen
NASA Goddard Space Flight Ctr., Greenbelt, MD 20771

H. Hojaji
The Catholic University of America, Vitreous State
Laboratory, Washington, D.C. 20064

K. B. Ma, M. Lamb, W. K. Chu
Texas Center For Superconductivity, University of
Houston, Houston, TX 77204-5932

ABSTRACT   Observing with robotic telescopes located on the Moon can become a reality in the near future (3 - 5 years). By using a combination of advanced ultra-lightweight optical technology, superconductors, radiation resistant detectors, and innovative launch strategies, it is possible to envision the deployment of multiple 1 m class fully steerable telescopes for UV-Vis-IR research, at a cost that can fit within the currently very tight budgetary constraints.

INTRODUCTION

My son Alan, who is fourteen and is here with me attending his first scientific symposium, asks me why there are only three papers in the Lunar Telescope section. My glib reply is "Son, this is an obvious case of quality over quantity". That of course is being totally facetious. I consider it a great honor to be addressing my peers who literally 'wrote the book' on robotic tele-scopes. The truth of the matter, let's face it, is that most astronomers think of lunar telescopes as a 'pie in the sky' concept. The project is too expensive, and there are too many unsurmountable technical problems. I hope you will permit me to present the case that the problem are surmountable, and that it is not a fantasy at all to think of observing with robotic telescopes on the Moon in the near future (3 - 5 years).

PROBLEMS OF BUILDING LUNAR TELESCOPES

There are many problems that must be solved before we can observe from the Moon. The major ones are :
1. Cost and financial resources
2. Weight
3. Telescope bearing and drive
4. radiation

Cost

At present the NASA budget is shrinking and Congress is debating the future of the Space Station, AXAF, CASSINI, etc. Prospects for the establishment of a lunar base, estimated to cost tens to hundreds of billions of dollars, is considered very unlikely in the current budgetary condition. Human settlement of the Moon is at least a decade or two away. By the same token, it is commonly thought that telescopes on the Moon, even remotely deployed and operated robotic units, will incur major expenditures in the range of hundreds of millions to billions of dollars. They are therefore equally unlikely to take place any time soon.

The allocation of public funds to a lunar astronomy project is an issue of national priority that cannot be addressed by scientists alone. Scientist can, however, take steps to make projects attractive by keeping costs to an absolute minimum without sacrificing scientific usefulness. We shall show how this can be accomplished.

Weight

We no longer have the Saturn V rockets of the Apollo era. There is currently in the U.S. a lack of heavy launchers capable of delivering tons of payload to the Moon. Table 1 lists present launch vehicles, their payload capacity, and estimated cost:

Table I Current U.S. Launchers

Vehicle	Payload Capacity[1]	Cost[2]
PEGASUS XL	5 - 10 (est.)	$ 15 M
TAURUS	10 - 15 (est.)	$ 30 M
DELTA 7925	350	$ 60 M
ATLAS IIAS/STAR 48B	700	$120 M
SHUTTLE/IUS	700	?
TITAN IV/CENTAUR	1750	$300 M

notes: 1. in kg, to lunar orbit
2. Sources: Pegasus, Taurus: Orbital Sciences Corp.; Delta, Atlas, Titan: Martin Marietta Aerospace Corp.

For comparison, consider the weight of some past and recent research spacecraft as listed in Table 2:

Table II Recent Astronomical Payloads

TELESCOPE/ SPACECRAFT	YEAR DEPLOYED	WEIGHT
LUNAR ORBITER	1967	390 kg
IUE	1978	670 kg
IRAS	1983	1020 kg
COBE	1989	4500 kg
GALILEO	1989	2500 kg
HST	1991	5500 kg

It is evident from the above tables that current low- and medium-sized (and -cost) launchers such as the *Pegasus* or the *Delta* are inadequate for the task of directly transporting most types of space telescopes to the Moon. Furthermore, unlike orbiting satellites, lunar telescopes require some type of landing mechanism plus support structures and drive mechanisms to function in a nonzero gravity field. The problem of limited payload capacity is therefore a serious obstacle. This problem is also of course related to cost, and we shall show how it can be overcome with the use of advanced technology.

Telescope Bearing And Drive
A telescope on the Moon must track 30 times slower with at least 10 times finer step sizes (0.05 arcsecond or better) than its earth counterpart. For a 1 m diameter gear, rotation through an angle of 0.05 arcsecond corresponds to a motion of 0.1 micron at the rim. At night the temperature of a radiatively cooled structure on the Moon is about 60 degrees Kelvin (McBrayer 1994), much colder than a cold night in the Midwest! Fellow observers who have worked through cold nights in the midwest know how the telescope and dome mechanisms tend to freeze, bind and break. Imagine how the gears will behave at liquid nitrogen temperatures! There are simply no mechanical bearing construction or lubricants that can function with such fineness in the lunar environment. This problem is well known to many scientists and has in the past led to suggestions of an early generation of fixed, nonsteerable 'transit' telescopes. We show that the bearing problem can be solved by innovative use of the recently discovered high temperature superconductors.

Radiation

The Moon has no atmosphere or magnetic field to shield against cosmic ray and solar flare radiation. The long term survival of electronics components and detectors are therefore of major concern. Previous suggestions for lunar telescopes almost invariably invoke the use of CCD (Charge-Coupled Devices, eg. Sykes et al 1990, Labeyrie and Mourard 1990) detectors. In fact one proposal for an early lunar telescope is totally dependent on CCDs for its operation (McGraw 1990, McBrayer 1994). We show that CCDs are not suitable for use on the Moon; better, hardier detectors are required for this application.

Fig. 1 shows the expected effect of solar flares on CCDs. The x-axis is the date, in this case the period October 1972 to February 1986. The y axis is the total fluence of protons with energy <1 MeV emitted during a flare. The horizontal dotted lines show the amount of damage on a 2048 x 2048, 25 micron square pixels after each event. The solar flare data is from Goswami et al (1988). The dotted lines are computed from the proton transfer equation of Janesick et al (1990) and are based on laboratory test data on CCDs for the CRAF/CASSINI project. It can be seen that even a relatively mild solar flare event can render an unshielded CCD inoperative. Over the period of time shown there were a total of 43 events - an average of about once every four months - that could have caused a CCD >99% charge loss. The damage is permanent and renders the multi-million dollar CCDs essentially useless.

Radiation damage in CCDs is a complex subject, but the basic physical mechanism is not difficult to explain. As an energetic particle hits the CCD substrate it dislodges silicon atoms, causing damage to the crystal lattice. A damage site becomes a charge trap or shows up as a hot spot. Since the operation of a CCD requires many 'bucket brigade' transfer steps across the device, the effect of even fairly mild local damages becomes cumulative and can exert an effect over the entire device. For example, a decrease of the CTE (charge transfer efficiency) to 0.999 translates into a 98.3% ($0.999^{4096} = 0.017$) loss in a 2048 x 2048 device. It is this dependence on almost perfect transfer of charge over a large number of pixels that makes a CCD so vulnerable to radiation. The bucket brigade mode of operation also means that, even when the radiation damage is local and is limited to a pixel or a small area, other areas of the CCD can be seriously affected. A bad pixel in a column can wipe out all information from upstream pixels in that column. If the damaged pixel is in a serial register it can render all upstream columns unusable and thus wipes out a whole area of the CCD. We can therefore see that it is the basic CCD structure and operation - namely charge transfer across a silicon substrate- that makes it so 'soft' to radiation, and why researchers are having such a hard time making rad hard

CCDs for space applications.

We need to develop hardier detectors that can withstand the high radiation level expected during long term operation on the Moon. At NASA Goddard we are developing an alternative type of array imagers known as Charge-Injection Devices, or CIDs, that can meet lunar radiation requirements.

OUR IDEA

Our idea is to use new materials and technology and innovative concepts. The major components are:
1. ultra-lightweight optics
2. superconductor bearings
3. charge-injection devices
4. advanced launch concepts

Ultra-Lightweight Replica Optics

The optics of a lunar telescope must be as lightweight as possible. The reasons are obvious. A mirror cell must support the primary mirror so as to maintain its shape. An optical tube must support the cell and the secondary optics and keep them in alignment. A mount and drive assembly must support the tube and point it, etc. Each element in the structure must weigh more than the element it supports in order to maintain a low center-of-mass. A reduction in the mass of the optics therefore translates into large reductions in the telescope and the payload, launcher, etc. This factor has been the driving force in space-based (and ground-based) telescopes over the past decades.

How do we make ultra-lightweight mirrors ? The traditional method of mirror making is by grinding and polishing a substrate, usually glass, and then coat it with a reflective metal film. This technique, shown in figure 2a, is limited by a minimum substrate thickness that is needed to avoid distortion by the work piece and printthrough of support structures. The lowest areal density that can be reached is about 20 kg/m^2 using beryllium.

We advocate using another method known as "replication". In this technique (fig. 2b) we start with a convex tool or mold. A piece of graphite "prepreg" or cloth is "layup" onto it, wetted with a resin, and cured. The shell takes on the shape of the tool and, after curing, becomes hardened and very stiff. It is then separated or "released" from the tool and coated with a reflective film. The replication technique is able to attain the minimum areal density possible because only as much material is used as is needed to maintain the optical figure.

The European Space Agency recently used replication to make optics for the XMM telescope, a grazing incidence x-ray telescope scheduled to be launched in 1999

(Egle et. al. 1990a,b). The ESA group have made pieces as large as 0.7 m O.D., 1 mm thick, with surface microroughness <2 nm. The achieved areal density is ~ 2 kg/m². Another group at CERGA, France applied this technique to make normal incidence optical mirrors with diameters of 0.5 m and greater (Assus and Glentzlin 1986). The mirror reproduced the shape of the tool down to an accuracy of $\lambda/50$ peak-to-valley (at λ = 6328 Å). They recently reported making a 1 m mirror with excellent surface quality. The viability of the replication technique is therefore well established and beyond doubt.

At NASA Goddard we have been experimenting with making ultra-lightweight replica optics over the past three years. We have successfully overcome a number of major problems including fiber print-through, release, and stability. Some of the results are shown in figure 3. We have also identified a class of new cyanate ester based resins which are superior to epoxy in terms of moisture absorption, dimensional stability, resistance to radiation, and ability to withstand repeated cycling between room temperature and immersion in LN_2.

Superconductor Magnetic Bearings
We have developed a new type of superconductor bearings for use in a lunar telescope. A basic bearing element (figure 4) consists of two opposing permanent magnets with circular symmetry with a YBCO (Yttrium Barium Copper Oxide) high temperature superconductor (HTS) sandwiched in between. The repulsive force between the magnets provide the lift against the load in the axial direction. A system consisting of only the two magnets is unstable, in accordance with Earnshaw's theorem. The insertion of a superconductor, however, imparts stability in both the axial and radial directions. This comes about by virtue of a unique property of the HTS known as "flux pinning". We shall give a demonstration and an explanation of this effect at the end of this talk. Our system of two magnets and a HTS is commonly referred to in the literature as a hybrid superconductor magnetic bearing or HSMB (McMichael et al 1991, 1992).

An HSMB is a non-contact system with typical gap between elements of a few millimeters. There is therefore no wear even for extended use at cryogenic temperatures. The system is passively stable in the sense that no power is required to maintain configuration as long as the HTS is in the superconducting state. On the Moon this condition is naturally satisfied. The night time temperature of a body in radiative equilibrium is ~60°K (McBrayer et al 1994), significantly below the transition temperature T_c~95°K of YBCO. Preliminary computations at JPL indicate that the HTS can probably also be maintained below T_c during daytime using a passive radiative cooling scheme similar to that suggested by Lin (1993).

A schematic of our HSMB or designed as the azimuthal support of an alt-az system is shown in figure 5a. A photo of the apparatus is shown in figure 5b. The HSMB weighs 7 kg and is designed to support a load up to 5 kg. It is fairly long (1 m) because of the anticipated need - which turned out to be unnecessary - for extra radial stiffness to accommodate heavy off-centered loading. Figure 6 shows our prototype ultra-lightweight replica telescope mounted on the HSMB. A more detailed description and analysis of the bearing mechanism is given in Ma et al (1994). We are in the process of designing a more compact equatorial mount mechanism.

Charge-Injection Devices
Instead of CCDs, we use intensified Charge-Injection Devices (I-CID) as detector for imaging and spectroscopy in the visible and uv. An intensified CID is a photon-counting detector with a microchannel plate (MCP) image intensifier optically coupled to a CID. The combination has the virtue of high dynamic range, high sensitivity, and is solar-blind in the uv version. The ICID is a hardier instrument and is much less affected by cosmic ray noise and solar flare radiation damage on the Moon.

The structure of a CID is shown in figure 7. Each pixel of a CID is composed of two MOS capacitors formed by the intersection of a column electrode and a row electrode. During an observation the photon-generated charge packets are accumulated under the column electrodes. During readout the charges are transferred to the row electrodes for sensing by an on chip amplifier, upon the selection of a row-column address. After readout the charges can be transferred back to the column capacitors for further accumulation (non-destructive readout). Alternatively the pixels can be cleared or injected, either individually or collectively, by collapsing the row and column potentials and causing the charges to be injected into the PMOS substrate.

CIDs can withstand high radiation levels because there is no transfer of charge (except within a pixel). A radiation hit will still induced a bulk displacement damage, but the damage remains localized and has no effect on the rest of the sensor. It is this lack of charge transport that makes the CID much less vulnerable than CCDs to radiation damage. As evidence of their hardness, CID imagers have been tested to 14 Mrad (Si) of gamma radiation (Carbone 1994) and 1 Mev equiv. neutron fluences of 10^{13} particles/cm^2 (Passenhelm and Golich 1992). These levels are more than two orders of magnitude higher the fataliity threshold of CCDs. CIDs are in fact routinely used in the inspection of interior of nuclear reactors.

CIDs are not new. They were invented at the General Electric Laboratory in 1972, at about the same time as CCDs (Michon 1972). CIDs have not received as much development as have CCDs and are therefore not competitive in commercial electronics (home video cameras, surveillance, etc). Recently however CID technology have received significant boosts from the SDIO (Strategic Defense Initiative Organization) for use in missile tracking, and from the Los Alamos National Laboratory (LANL) for monitoring high speed bomb tests. Both those technologies have become available for civilian use. Our group at NASA Goddard are working with the manufacturer under a NASA grant and an SBIR (Small Business Innovative Research) program to develop advanced CIDs for space UV astronomy. Work is also currently under way to modify the design of a flight qualified, transputer-based controller system for a CID. The controller was originally built to operate a photon-counting CCD camera for the SOHO/CDS project.

Advanced Launch Trajectories And Mission Concepts
Using replication technology, we estimate that a 1 m telescope, including optics, structure, detectors and electronics, will have a mass of ~30-40 kg. Belbruno (1994) has computed that a Peagsus XL can deliver this size package to lunar orbit using a trajectory based on the "Weak Stability Boundary" (WSB) technique. A Pegasus, as some of you may recall, is a small rocket launched from an aircraft. At a per unit cost of ~$10M, it is just about the cheapest launch platform in existence. The WSB technique makes use of the fact that, in the Earth-Moon-Sun system, there is a region ~10^6 miles from earth that is at infinity as far as the Earth's and the Moon's gravity field is concerned. A launch to this region, follow by a small orbital manoeuver of ~30 m/s, causes a spacecraft to go from the Earth's gravity field into the Moon's gravity field. Another small manoeuver causes the injection into a circular orbit around the Moon. The process takes about three months instead of the three days required by a direct or Hohmann type trajectory. The extra time is not a concern for an unmanned mission. The validity of the WSB technique has been demonstrated in the case of the Japanese *Hiten* spacecraft (Belbruno 1994).

CONCLUSION

We are still in the process of working out many other details, such as how to package, soft land and robotically deploy a telescope. The overall picture is becoming more focused, however. By a refinement of proven advanced technology, it is possible to make lunar telescopes that are very light, radiation-resistant, and can point and track with precision. By using innovative launch concepts it is possible to keep costs down to the level of $10-20M per mission. With the use of these advanced materials and techniques, lunar telescopes are no longer in the 'pie in the sky' category. It is now possible to envision the deployment of many small expendable launches over a period of time to gradually build up a full scale lunar observatory. The observatory will consist of many telescopes gathering data over the entire electromagnetic spectrum from low frequency radio through IR, visible, UV and up to gamma rays. Equipped with state-of-the-art instrumentation and making full use of the fantastic conditions on the Moon, these new generation of telescopes promise to produce a scientific bounty that is beyond anything we can imagine.

Thank you for your attention.

ACKNOWLEDGEMENTS

We gratefully the help of many colleagues whose contribution have made this work possible. We especially thank C. Frum, D. Ziobro, P. Perry, B. Turnrose, R. Pitts, P. Pitts (CSC), J. Baker, C. Condor, J. Houston, D. Linard II, N. Jenson, S. Ollendorf, K. Segal (NASA/GSFC), R. Davis, C. Lodge, M. Wilson (NASA/KSC), C. Ortiz, C. Schomberg, J. Christiansen (NASA/JSC), R. Churchward (NASA Ames), R. Carmichael (NSO), and G. Downing, H. Chen (NIST).

We acknowledge helpful discussions with J. Carbone (CIDTEC, Inc.), Y. Flom (NASA/GSFC), J. Hull (ANL) F. Moon (Cornell), H. Murakami (ISTEC, Japan), J. Peters (Martin Marietta), S. Shore (SBU), J. Tompkins (NASA/LaRC), and G. Yates (LANL).

CID detector development is supported under NASA Research Grant RTOP-188-41-24-41, Dr. R. Kimble, PI.

REFERENCES

Assus, P. and Glentzlin, A. 1986, Proceedings SPIE 628, Advanced Technology Optical Telescopes III, p. 545.
Belbruno, E. 1994, J. Brit. Interplanet. Soc. 47, p. 73.
Carbone, J. 1994, Private Communication.
Egle, W., Bulla, H., Kaufmann, B., Aschenbach, B. and Brauninger, H. 1990, Optical Engineering, 29, p. 1267.
Egle, W., Bulla, H., Scheulen, D., Aschenbach and B., Brauninger, H. 1990, Optical Engineering, 29, p. 1260.
Goswami, J.N., McGuire, R.E., Reedy, R.C., Lal, D. and Jha, R. 1988, JGR, 93, p. 7195.
Janesick, J., Soli, G., Elliott, T. and Collins, S. 1990, IEEE 27th International Nuclear & Space Radiation Conference, Reno, Nev. July 1990.
Labeyrie, A. and Mourard, D. 1990 in Astrophysics from the Moon, AIP Conf. Proc. 207, eds. M.J. Mumma and

H.J. Smith, p. 538.
Lin, E.I. 1993, *NASA Tech Briefs*, Sept 1993, p. 82.
Ma, K.B., Chen, P.C., Lamb, M. 1994, in press.
McBrayer, R.O. et al 1993, <u>Lunar Ultraviolet Telescope Experiment (LUTE) Interim Technical Assessment</u>, NASA Marshall Space Flight Center, March 12, 1993.
McBrayer, R.O. et al. 1994, <u>Lunar Ultraviolet Telescope Experiment (LUTE) Phase A Final Report</u>, NASA Tech Mem. 4594, NASA Marshall Space Flight Center, April 1994.
McGraw, J. 1990, in <u>Astrophysics From the Moon</u>, A.I.P. Conf. Proc. **207**, Annapolis, MD, eds. M.J. Mumma and H.J. Smith, p. 433.
McMichael, C.K., Ma, K.B., Lin, M.W., Lamb, M.A., Meng, R.L., Xue, Y.Y., Hor, P.H. and Chu, W.K. 1991, *App. Phys. Lett.* <u>59</u>(19), p. 2442
McMichael, C.K., Ma, K.B., Lamb, M.A., Lin, M.W., Chow, L., Meng, R.L., Hor, P.H. and Chu, W.K. 1992, *App. Phys. Lett.* <u>60</u>(15), p. 1893.
Sykes, M.V., Vilas, F., Page, T.L., Smith, H.J., Burns, J.O., Colavita, M., Snyder, G., Stern, S.A., and Talent, D.L. 1990, in <u>Astrophysics From the Moon</u>, AIP Conf. Proc. **207**, eds. M. J. Mumma and H. J. Smith, p. 328.

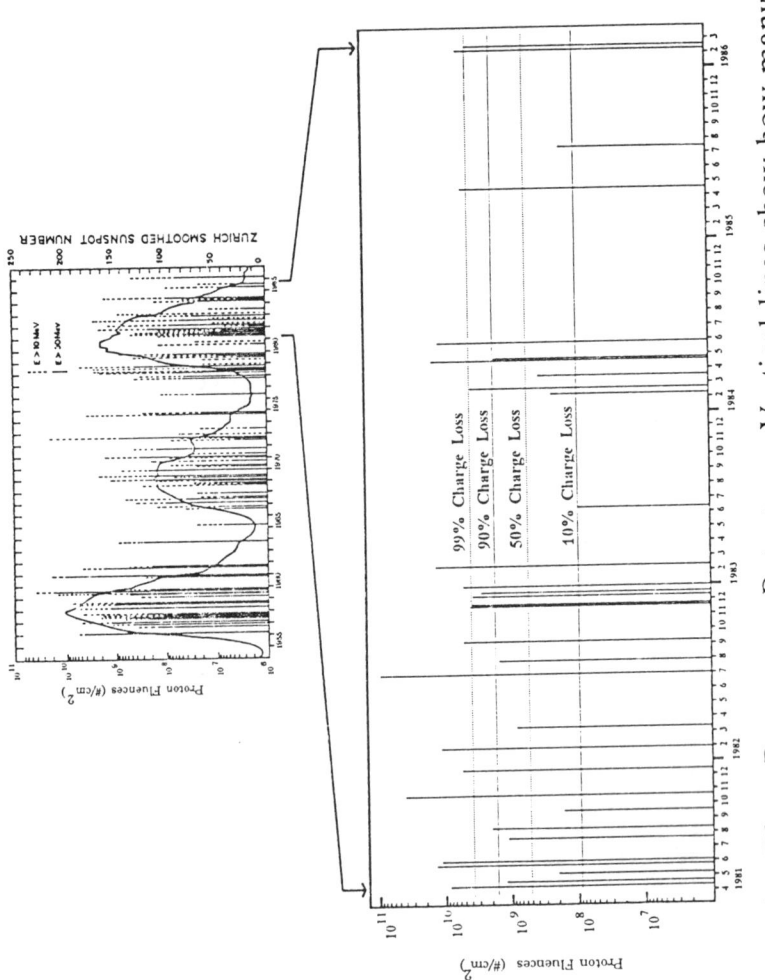

Solar Flare Damage to Detectors: Vertical lines show how many particles were in a solar flare. Horizontal dotted lines show the damage to a CCD after a *single* flareup.

ROBOTIC TELESCOPE FOR THE MOON

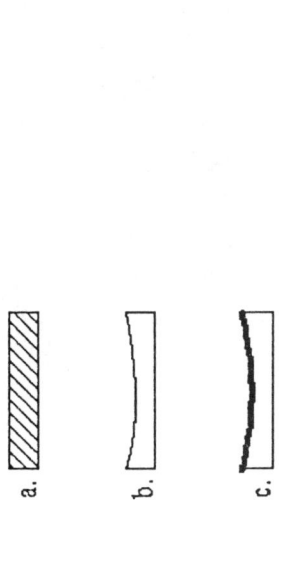

Traditional Technique

The traditional way of making mirrors:
a) start with a piece of substrate, usually glass.
b) grind it to shape.
c) overcoat with reflective film.

This technique is limited by the need for a minimum substrate thickness for grinding and to prevent print-through. Using the lightest material, beryllium, the achievable areal density is about 20 kg/m**2. As an example, areal density of the *Hubble Space Telescope* primary mirror is about 180 kg/m**2.

Replication

a) A mandrel is first polished to shape.
b) A piece of prepreg is applied, coated with epoxy, then cured to make a shell.
c) The graphite epoxy shell is released from the mandrel.
d) The shell is overcoated with reflective film.

The shell only needs to be as thick as is required to maintain its figure under operational conditions. As an example, the European Space Agency's XMM project has succeeded in making replica optics with thickness of ~1 mm and areal density of ~2 kg/m**2.

Optics & Structures Made With Advanced Composite Materials

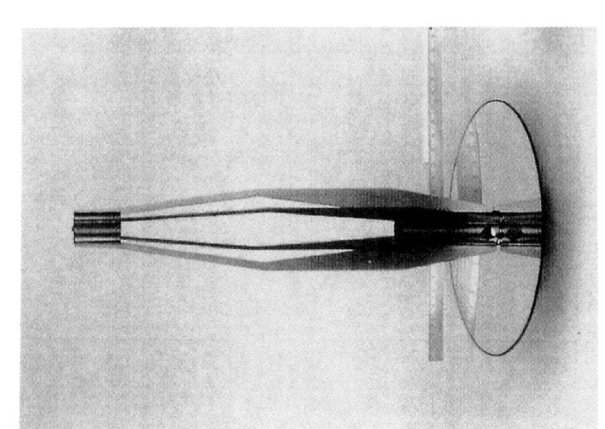

c. An All-Graphite Epoxy Telescope. A Meinel-type telescope with 42 cm f/4 primary mirror. The primary, secondary, and tertiary mirrors and the support structure are all made from graphite epoxy. Total weight 1.2 kg.

a. Two Graphite-Epoxt Mirrors Made by Replication. Left: a 42 cm (16.5") f/4 Cassegrain primary, 5mm thick graphite epoxy, weight 830g. Right: 20 cm f/0.6 mirror, 5 mm thick graphite epoxy.

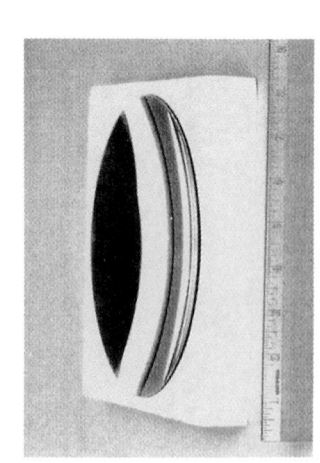

b. The 20 cm f/0.6 in a Mirror Cell. The cell was machined at NASA KSC from a piece of Space Shuttle heat tile material.

Superconductor Bearings

We suggest using *High Temperature Superconductor Bearings* for telescope drives on the Moon. HTS bearings are non-contact and therefore do not wear out. The motion of the bearings can be precisely controlled by coils and electronics circuitry. No active cooling of the HTS is required on the Moon; natural radiative cooling is adequate to keep the HTS in the superconducting state.

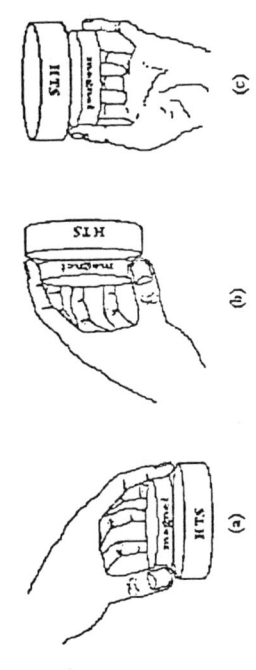

The recently discovered high temperature superconductors (HTS) exhibit a unique phenomenon known as 'flux pinning'. In fig. (a), a magnet is held above a YBCO disk while the latter is being cooled by liquid nitrogen to the superconducting state. The magnet and the HTS becomes attached to each other without physical contact. In fig. (b), when the magnet is raised and turned vertical, the HTS follows along. In fig. (c), the HTS maintains its position even when the magnet is held upside down. The magnet and HTS can rotate freely with respect to each other by virtue of symmetry of the magnetic flux distribution. This is the basis of our superconductor telescope bearing.

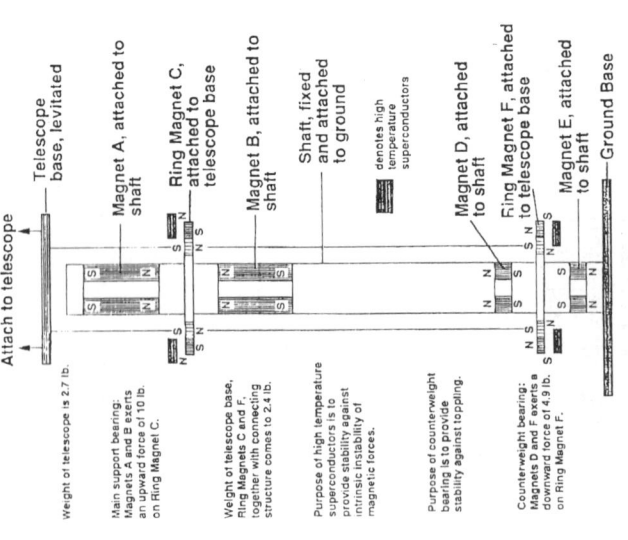

A proof-of-concept superconductor bearing built by scientists at the Texas Center for Superconductivity at the University of Houston. *Left*: schematic and *Right*: photo of unit. Opposing magnet pairs are used to support the load while the HTS (dark areas) 'pin' the magnets to provide stability. The system is passive in that no power is required for the bearing to maintain its configuration. The unit is 81 cm tall and 56 cm wide (32"x22") at the base. It is designed to support a load of 5 kg (10 lbs) with extra stiffness in both the radial

A demonstration of our ultra-lightweight, all-graphite-epoxy telescope mounted on a superconductor bearing given at a meeting of the *American Astronomical Society* held in Washington, D.C. January 1994. The telescope has a 42 cm (16.5 inch) diameter primary mirror and a 5 cm (2 inch) secondary mirror. It measures 60 cm (24") high and weighs 1.2 kg (2 lbs 10 oz). The superconductor bearing is 81 cm (32 inches) high, 56 cm wide (22") at the base, and weighs 7 kg (15 lbs). Photo by Britt Griswold.

Charge-Injection Devices (CIDs)

A new type of solid-state detectors is being developed at NASA Goddard (R. Kimble, PI) for space astronomy. CIDs are silicon-based imaging arrays similar to CCDs. Unlike CCDs, they do not transfer charge and can withstand much higher levels (100-1000x) of radiation.

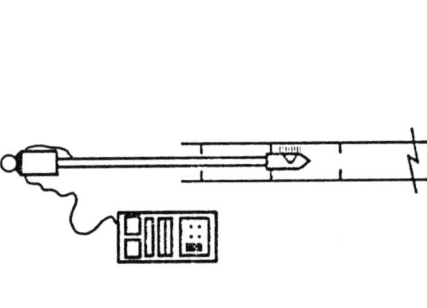

CIDs are so radiation resistant that they are used for inspecting the interiors of nuclear reactors. This tool is made by Image Technologie of France, using a sensor made by CIDTEC Inc. of Liverpool, N.Y.

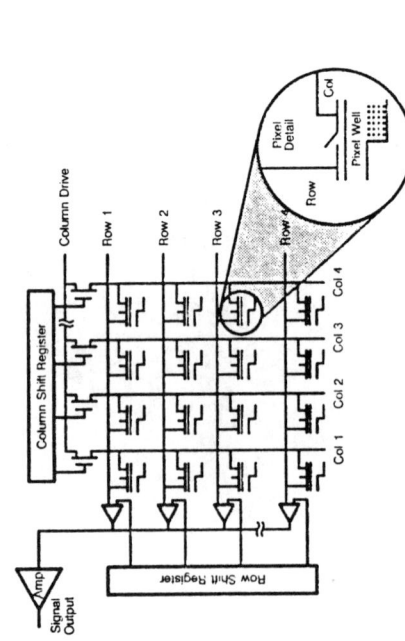

Charge-Injection Device (CID) architecture. Each pixel is defined by the intersection of a column and a row electrode. Individual pixels can be randomly accessed for signal readout or reset operations. Unlike in a CCD, there is no charge transfer across the entire sensing area during readout. Radiation damage in a CID pixel is therefore localized and does not affect other pixels.

D516.012

AUTHOR INDEX

A

Adelman, S. J.: 20
Au, W.: 17

B

Benkhaldoun, Z.: 239
Brecher, K.: 93
Bresina, J.: 101
Burns, J. O.: **242**

C

Chen, P. C.: **252**
Chu, W. K.: 252
Cornell, M. E.: 136
Crawford, D. L.: **223**

D

Donahue, M. A.: 129
Drascher, E.: 101
Drummond, M.: **101**
Dukes, R. J.: **20**

E

Eaton, J. A.: **226**
Edgington, W.: 101

F

Filippenko, A. V.: 86

G

Graves, M.: 167
Guinan, E. F.: 20

H

Hall, D. S.: **65**
Hansson, O.: 148
Hearnshaw, J. B.: **233**
Henry, G. W.: **37**, **44**, 101
Hereld, M.: 195

Hojaji, H.: 252

I

Iverson, D. L.: 120

K

Kadiri, S.: 239
Kelton, P. W.: **136**
Kubinec, W. R.: 20

L

Lamb, M.: 252
Leiker, P. S.: **93**
Lynch, J.: **205**

M

Ma, K. B.: 252
Malina, R. F.: 184
Markworth, N. L.: **81**
McCollum, B.: **167**
McCook, G. P.: 20
Miller, G. E.: **173**
Mohan, S.: 148
Monahan, C. M.: **120**
Morgan, T.: **184**
Muscettola, N.: **148**

N

Nations, H. L.: 20
Nguyen, H.: 195

O

Oliversen, R. J.: 252

P

Paik, Y.: 86
Patterson-Hine, F. A.: 120
Pell, B.: 148
Percy, J. R.: **17**

Q

Querci, F. R.: **239**
Querci, M.: 239

R

Rauscher, B. J.: **195**
Richmond, M. W.: 86

S

Sadler, P. M.: 93
Seeds, M. A.: **11**
Severson, S.: 195
Smith, D. P.: 20
Swanson, K.: 101

T

Treffers, R. R.: **86**
Trimble, V.:

V

Valentine, K. M.: **129**
Valentine, R. W.: 129
Van Dyk, S. D.: 86